Ubiquitous Voice: Essays from the Field

Edited by Lisa Falkson

For my father, Eric Falkson

TABLE OF CONTENTS

Introduction

By Lisa Falkson

Why this book, why now?

In the 1990's and early 2000's, speech recognition technology was limited to telephony applications, mainly replacing old DTMF systems or live agents with speech automation. At that time, no one would have taken the time to read a collection of essays like this one – nor would we have had the motivation or material to write it.

As a designer of IVR speech recognition applications early in my career, telling someone my occupation at a cocktail party was a fatal mistake; I was immediately placed in the same category as a tax auditor – someone who made people's life more difficult. "I hate those speech recognition systems" and "Why does that stuff never work?" everyone asked.

Luckily, times have changed. Hardware and software advancements have nearly perfected recognition accuracy, smartphones were invented, and we witnessed the arrival of the smart home speaker. Quite suddenly, an old industry has become new again, and reached the mainstream.

A growth industry

The past few years have seen rapid rise and popularity of speech recognition and natural language systems. From Siri in your pocket to Alexa or Google Home on your countertop, the proliferation of devices has made these systems more commonplace. In particular, the smart speaker space has exploded, with roughly 40 million total devices sold in the United States.

With the market expanding, there becomes a need for a new, larger population of speech experts to design the future of these devices and applications. This was, in fact, the original motivation for this book. We hope to share a real-life view of the world of design, development, and product management of speech products and applications. We also include a market overview, as well as some history and personal stories from key contributors in the space.

Purpose of this book

The goal of these essays is to provide perspectives from various experts in the field. From designers to product managers to analysts to CEOs, they each share a unique perspective on an aspect of the speech industry. From the complexities of adding voice to hardware to the challenges of maintaining user privacy in speech systems, our authors tackle some of the most difficult subjects in the industry today. A few authors (including myself) discuss the use of speech technology in key verticals, such as healthcare and automotive. Others take a look at the importance of design: pondering the depth of personality required in a speech system, strategies for making a VUI more human, and the need for liberal arts experts to participate in quality design. Some of the more technical chapters review the obstacles to designing speech applications and the next steps in NLU. There are also personal stories in this book: experiences working on Echo/Alexa, and the development of an early smart speaker competitor, ivee.

The combination of different perspectives should appeal to anyone interested in the speech technology industry, whether novice or expert. This book aims not only to educate, but also to spark future thought and discussion. Hopefully, the reader will finish reading with enhanced knowledge and passion for a growing field.

Reclaiming our Moments

By Sunjay Pandey

Every week during orientation in the heart of Amazon's campus in Seattle, each new employee learns an internal slogan built with three imperatives: "Work hard, have fun, make history[1]." In eight out of twelve Pacific Northwest months, a gray landscape hangs outside of the plate glass windows of orientation classrooms. Still, these six words tend to stoke the red-orange fires burning in the wild-eyed builders, entrepreneurs, and type-A's Amazon tends to attract and select. It did, too, for me.

I have been removed now from Amazon and our launch of the Alexa Skills Kit (ASK) for more than two years. I've met with talented people who have wondered about how to plan for and find a job at Amazon. To a person, I encourage them to find roles at the company that will allow them to satisfy at least two components of the slogan: "have fun, make history." But, "making history" is no walk in the park; it requires sustained, extreme effort. So, to make the most of their time at Amazon – I confirm for the ambitious – they should seek roles where they will satisfy the tribal slogan in whole. This is precisely the type of role I held in May of 2015 as Head of Product, in a plain, five-story building named "Prime" in the South Lake Union district of Seattle, preparing to ship ASK alongside a band of voice technology missionaries.

---

[1] The slogan is a derivation and evolution of a quote from Jeff Bezos' original letter to shareholders in 1997: "It's not easy to work here (when I interview people I tell them, 'You can work long, hard, or smart, but at Amazon.com you can't choose two out of three')."

Summer shines sharply in my memories of working on Alexa. My family moved to Seattle in 2012 for my role at Amazon Web Services (AWS). In July of 2014, after living the three-part Amazon motto in AWS, I decided that I wanted to explore the Amazon outside of the runaway cloud computing locomotive. On the internal job site, I found a job description that sounded intriguing if vague. It was a role in "mobile." The hiring manager, John, invited me to dinner after a first meeting. He pitched me on the Head of Product role for a device Software Development Kit (SDK). Over the next two weeks, I had a series of interviews with other members of the team about a secret project (called "Doppler" or Project D"). By August, I was using a test Echo unit in our hot, air-condition-less rental home. Echo (and Alexa) had no SDK yet. Still, I wished I could ask Alexa to make it cool. Instead, I played music, set timers, asked all sorts of questions and took care to prevent anyone outside of my family from seeing the black cylinder or hearing Alexa. My mind filled and contorted with the possibilities of ambient computing accessed by a user's voice, hands free. I would be defining the product strategy for Alexa's SDK - working on a team that would empower people to speak aloud in their homes so that an intelligent personality resident in the cloud would carry out their requests. I had spent nearly 20 years building technology products. This was the most meaningful opportunity and groundbreaking technology I had ever been part of – by a moonshot.

I am grateful for having had the chance to work with amazingly visionary and driven colleagues at Amazon. I count myself lucky for being able to "think big" about what the Alexa Skills Kit could help the world create. In the same way that the Mosaic browser and its point-and-click interface is credited with making the World Wide Web accessible and comprehensible to humanity, voice-based assistants are the front-runners to – and accelerators of – ubiquitous computing where all manner of "things" will be connected, controllable, and changeable via the Internet. The marketing buzzword for this emerging future is the Internet of Things (IoT). People will consult, check, control, change, and be notified by these things. In current state, consumers will use traditional apps to interact with IoT. Increasingly though, human beings will use their voice to talk to and hear from connected things. The birth of Alexa (and other voice assistants) will make this IoT future come to life sooner and more explosively in much the same way that Mosaic led to an explosion in the growth of the web (and web pages). This explosive growth will eventually lead to companies increasing the supply of available devices across a range of uses, reducing the costs of IoT and making it affordable for the mainstream.

Stating the obvious, a future where you use your voice to interact with an ambient computing environment will yield advancements that are important to society. There are educators, founders, politicians, and professionals of all stripes who will make certain of this. But, as people adopt hands-free voice assistants in their early days, it is the compounding accumulation of happiness from simplifying "little" everyday tasks and automating the mundane that will compel consumers to seek ubiquitous voice interfaces.

That same August of 2014, besides wishing Alexa could cool our house, I wished she could water the lawn – mainly to satisfy our zealous landlord and her twice-a-day watering regime. Our rental home had a sprinkler system – supposedly convenient, automated and "smart." After two visits to the house herself that summer of 2014 to fix her system, our landlord called in the maintenance pros. The technicians came and went twice. The call from our landlord and the sticky note the technician left on the sprinkler control unit came across like cuneiform. With two kids under 5, long work days, continual learning (or more work) and minimal time to unwind in the evenings with my wife Ellen, I could think of better ways to invest my time than solving a lawn-watering problem. I loudly said, "Alexa, please water the lawn!"

In this state of temporary (and "First World") frustration, I deeply internalized the possibility of Alexa and the Skills Kit – and more broadly ubiquitous voice technology. Voice tech is not simply about convenience and ease. Instead, ubiquitous voice enables us to reclaim bits of our modern lives (and time) that have been smothered like forests in the Southern United States have been overgrown by Kudzu. The "invasive weed" of our lives though is many-faced. It's the mobile apps that take our presence and attention from our loved ones; the lack of time from a dual-career home that most Americans lead to provide for their families; the lack of time from having to take care of household chores, the dissipation of extended families from neighborhoods and towns as they seek work so that it is physically impossible for one family member to lend a helping hand to another family in another geography.

Hyperbolic? Perhaps. Imagine though the bleary-eyed new mother who asks Alexa to set a timer so she can measure how long she's breastfed her newborn as she works hard to make sure her breast milk comes in. Imagine the mom who asks Alexa to sweep the floors, gaining twenty minutes back at night to her daughter with her math homework. Imagine the father who asks Alexa to playback video from the front porch at the exact moment that the home security system sounded its alarm. Imagine the daughter who asks Alexa to rewind the most difficult steps of a ballet – again and again - as she works to perfect her routine. Imagine the son who asks Alexa to lower the basketball rim to 8ft so that he can practice his jump shot without straining his form. Imagine the grandfather, who asks Alexa for the date of a Revolutionary War battle as he seeks to tell a more complete family history to his granddaughter. There are hundreds of daily moments like these that will be powered by ubiquitous voice technology. Taken together, these reclaimed moments can help us lead lives with greater humanity and meaning because we can invest more time and consciousness with the people and avocations we cherish.

Between September of 2014 and May of 2015, I led Product Management for the Alexa Skills Kit. Our team worked to define and build an SDK that could help "voice app" developers build an Alexa Skill within minutes. As we prepared to launch, in the Amazon working backwards way, I crafted an imaginary, internal press release that communicated a future for customers. One of the use cases I described was for lawn watering. "Alexa, ask my sprinkler to water my lawn for 15 minutes" found its way into our public launch materials on June 25th 2015. That same day on the Rachio Sprinkler Community forum, a user with the handle *wbhartmanii* wrote that he had signed up for the Skills Kit, shared the link to sign up, and said he would begin working on an Alexa integration. By June 29th, *wbhartmanii* wrote: *Just said "hey Alexa, tell the sprinklers to water the back yard"....and she did ☺ Thanks Alexa!* By August, Rachio had announced they were working on an official solution. At the time of this writing, developers have launched 30,000 Alexa skills – with the pace accelerating. The movement to reclaim human moments is on.

Because of my time at Amazon, I admit that I am biased. Though, I don't think it's hyperbolic to say that the launch of Echo/Alexa in 2015 will mark the tipping point in society's appetite for conversational interfaces. Voice technology will become ubiquitous and benefit from: democratized artificial intelligence (see Google's open sourcing of its Tensor Flow machine intelligence technology), variable computing costs bending towards zero (cloud and fog computing), and a deep-seated human need to simplify daily life and find time to re-humanize through deeper personal relationships. As a result, the acceleration of voice technology will hasten the Internet of Things. While the automation of work that was previously done by people will continue to be an existential issue for our national identity and our economy, new entrepreneurs and industries will be formed. Ubiquitous voice will inspire and require new experiences that are multi-modal – engaging more than just the sense of hearing. Mixed reality applications where a user can use gross and fine motor control along with voice response to interact with their environments is already being tested and will be available soon. And while ubiquitous voice as the state of the art in interfaces will be transient, the potential of its value in our lives is the freedom it affords us – freedom stitched from time reclaimed from the mundane so that we might have more moments that matter with the people and experiences that sustain us.

Obstacles to Ubiquitous Speech Interfaces

By Bruce Balentine

Abstract

There is great enthusiasm today that speech user interfaces are becoming effective, efficient and satisfying—so much so that an increased rate of diffusion into diverse market niches is impending. "Soon, speech will be in constant use everywhere," we are told. However, no discussion is complete without considering the obstacles that stand between the promise of ubiquitous speech and its realization. In this article, I discuss specific limitations of design, technology, integration and architecture that hold back the explosive growth of products. Such limitations include pesky and neglected issues of interactivity, including turn taking, error-recovery, and reusable speech-behavior primitives.

Introduction

Speech and voice technologies segment into three functional categories: speech recognition, speech synthesis, and voice biometrics. All share a common research background, and all play a part in the emergence of Artificial Conversational Speech (ACS) applications. While ACS has evolved for many years, various tracks have matured at different rates, shown Figure 1.

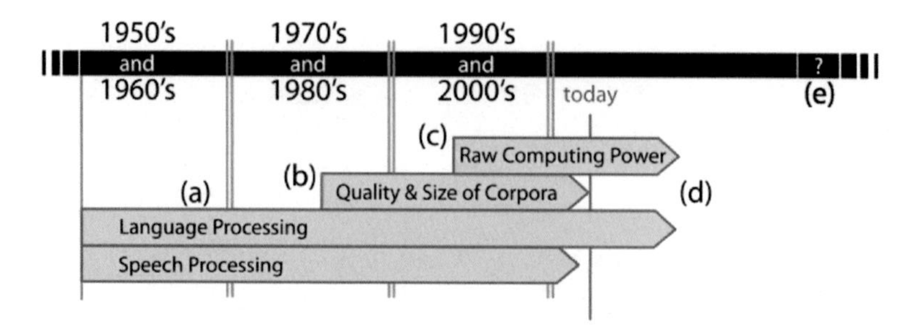

Figure 1 - Progress in ACS is deceptive, because it is measured in terms of distance from the origin rather than distance to the goal.

Starting with early work at Bell Labs and elsewhere, basic speech and language processing (a) have advanced significantly—to the point of nearing full maturity. Massive and high-quality databases (b) are now collected, and raw computing power (c) continues along its predictable trajectory. This timeline implies that we are nearing a technological tipping point (d), a presupposition for its ubiquitous uptake (e), presumably in the near future.

Despite this historical progress, the prediction of incipient ACS ubiquity overlooks several additional tracks of some complexity. These tracks include the 'theory of mind' (ToM) and its dependence on consciousness (Balentine 2007), the biological basis for human emotions (Damasio 1999), the fact that language is an evolved structure, deeply specific to the human condition (Pinker 1994), and several ethical and social obstacles of unpredictable impact (Christian 2011). Serious work on these tracks has barely begun and cannot be plotted. Under such circumstances, it seems appropriate for a devil's advocate to point out the more glaring of these historically persistent obstacles.[2]

---

[2] Devil's advocates are those who adopt a position contrary to an accepted norm—and even contrary to their own beliefs—for the sake of debate. See https://en.wikipedia.org/wiki/Devil%27s_advocate for more on devil's advocate argumentation.

Product Diffusion and Ubiquity

New technologies tend to diffuse through a user population according to predictable patterns described by diffusion theories [Rogers 1962, Bass 1969, Moore 1991].

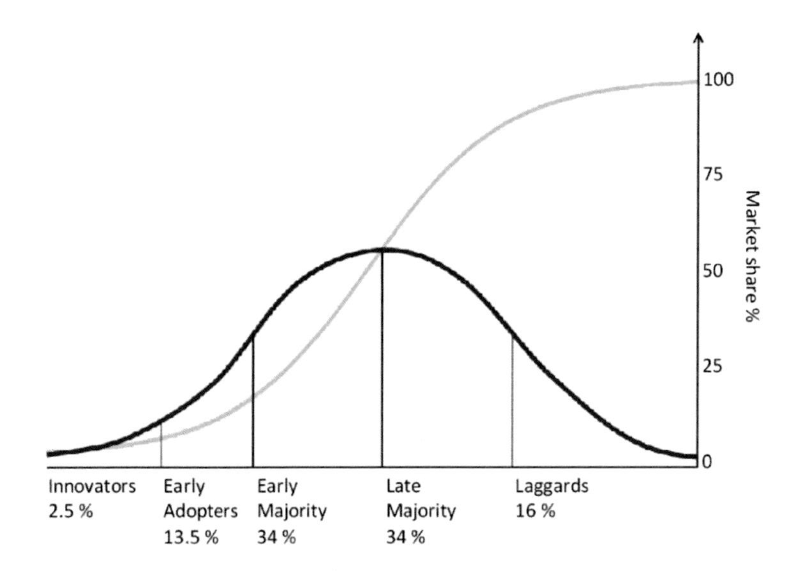

Figure 2—The famous S-curve by Everett Rogers showing the diffusion patterns of innovative ideas and products

For decades, the speech industry has tended to underperform against this model, for three reasons: pundits overestimate 1) the intrinsic value of the technology, 2) the rate of progress toward its commercial success, and 3) the willingness of end users to overlook disparities between product claims and reality. This overestimation is exacerbated by underestimation of the usability, interaction design, and engineering fundamentals that erect obstacles to product usability. The result has been that new speech products are embraced by early adopters, but refuse to go mainstream. The one exception is IVR, a ubiquitous technology that is imposed on users without their choice, and which is a cost burden to the buyer rather than a profit-generating product that benefits the user. If we define "ubiquitous" as the transition in uptake from the early majority to the late majority—

the middle of the bell curve shown in **Figure 2**—then we must do much better than that.

One-Shot Recognition Versus Interactivity

The first and most glaring of obstacles derives from unsolved issues of interactivity and their avoidance through one-shot dialogues (**Figure 3**). In this model, a human user initiates a conversation by speaking a phonetically- and grammatically-complex utterance—not a word or two to launch an interactive conversation, but a single sentence that contains all relevant information. The application then relies on so-called Natural Language (NL) speech recognition to capture, frame, parse, and interpret the phrase, counting on grammatical patterns, natural prosodic fluencies, and redundancies in the input to arrive at reliable actionable user intentions. NL technology is mature, so designers and developers have migrated instinctively to this "one-shot" model of conversation.

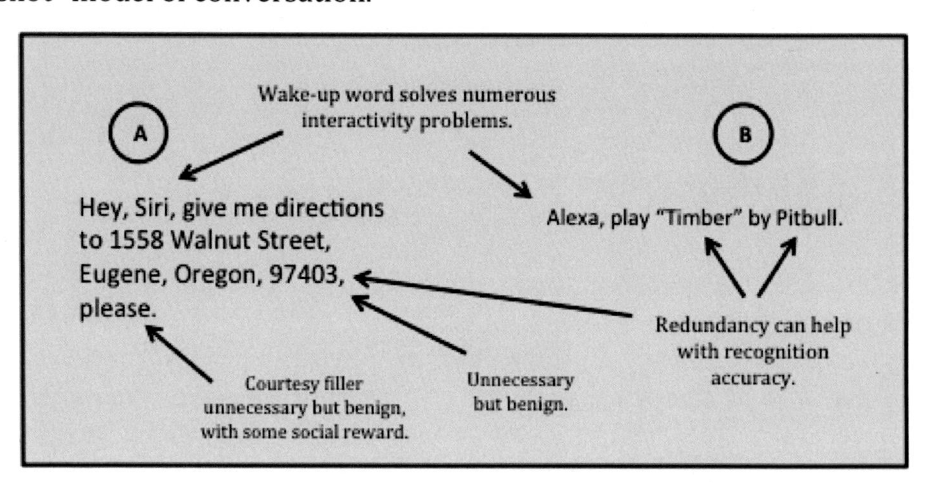

Figure 3—One-shot utterances provide conversational cover, concealing the absence of interactivity.

Such utterances are fine when the user possesses knowledge required to construct them. We see great success, therefore, with GPS-location services (**Figure 3A**), music-selection behaviors (**Figure 3B**), casual search tasks and simple transcription (such as voice-texting). The fact that a number of UI features are extremely unnatural—the requirement in most cases for an attention word, the need to cram all data into a single breath (turn), the intolerance for flubs and stutters, and the difficulty of self-correcting errors or changing one's mind—is covered by the fluid and grammatical tolerance allowed by the one-shot. In other words, it's good enough to do useful things.

But most complex user goals require interactivity in the true sense—the user does not possess all required knowledge, but instead is searching for a solution to a complex (and sometimes vague) problem. In such cases, it is impossible to construct a one-shot utterance, because the task fundamentally requires conversational give-and-take. This is true of all machine-initiated and mixed-initiative dialogues, including negotiations, form-filling, inspection, command-and-control, and similar multi-turn conversations.

Turn Taking

One of the foundational elements for interactivity is turn taking (Attwater & Balentine, 2009). When two humans converse, there are basic rules in play that govern which interlocutor currently has the floor, as well as when and how the turn might be yielded to the other. These rules are both explicit—grounded in social power, deference to mutual knowledge, and established communication protocols—and implicit, with unconscious emotional underpinnings that dominate the interaction. Turn taking thus has a strong externally observable rhythmic and temporal component, as well as a hidden internal component that is driven by psychological forces such as personal urge and theory of mind.

Users naturally transfer both innate and learned turn-taking protocols from human-human conversation to human-machine conversation. This natural transference is almost always detrimental to usability, as users bring invalid assumptions to the effort, including assumptions about mutual knowledge and social power, the nature of machine intelligence, and belief in the marketing claims of the product developers. The result is a consistent obstacle to user success, acceptance, enthusiastic word-of-mouth recommendation, and purchasing behaviors that are critical in technology diffusion theories. Inattention to turn taking has been a significant obstacle to successful speech deployments since the inception of the technologies. The questions, "When do I speak, and when do I listen? Can I interrupt? How do I get its attention when things go wrong?" continue to be a leading source of anxiety and error on the part of human users. Many attempts at a solution—e.g., half-duplex interactions with prompting tones or push-to-talk buttons; full-duplex interactions based on barge-in or independent talking and listening channels; and one-shot utterances with attention words ("Hey, Siri" or "Alexa")—have been floated with little knowledge of or interest in the underlying psychology that dominates human turn-taking behaviors.[3]

A more intelligent and far-reaching general-purpose solution to the turn-taking challenge has been available for at least 15 years (Ström & Seneff 2000), and a formalized protocol that allows humans and machines to exchange spoken information reliably and comfortably should be in place by now. The externally-observable (explicit) rules are easily reduced to algorithmic control, and the basics of a management engine capable of knowing whose turn it is to speak, and what to do in the case of interruptions, is well within reach (Attwater & Balentine, 2010).

Short-Term Memory and Theory of Mind

I recently observed a user who, while driving, had noticed that Siri seemed to be unresponsive to his recent request for directions. Just as he was exiting the parking lot, I heard this quick exchange:

[3] This is especially odd given that the subject is very well and competently researched. Solutions are well-known, and fairly well-tested.

Driver: Hey, Siri, how are you coming with those directions?

Siri: Directions to where?

Dialogue 1—Lack of memory creates amusing interactions.

One of the biggest advantages of a one-shot dialogue is the avoidance of the need for short-term memory. But without it, an interactive dialogue stumbles quickly. It's very difficult to maintain a conversation when one of the conversational partners fails to remember what was recently spoken. But keeping up with the steps of a conversation requires more than mere memory—it requires an effective model of self and other. In his paper on speech communication with robots, Roger K. Moore discusses both memory and intention:

> If a user takes a *design* stance to an object or device, then any unexpected behavior is taken to indicate that the device is broken and interaction should be abandoned. However, if a user takes an *intentional* stance to a device (which is highly likely for a robot), then any unexpected behavior is taken as evidence that there are hidden motivations and goals that need to be determined and perhaps changed (Moore 2015).

The ability to model the partner in a conversation depends on theory of mind (ToM). When two humans converse, each is maintaining a theory of the other's mind—that person's intent, goals, emotions and conversational context. Humans are very good at this. Humans adopt the same assumptions when conversing with machines, forming a theory of the machine's mind and making inferences about its capabilities and intentions. Indeed, the user might be said to be *probing* the machine's mind. But invariably, such probing exposes the machine's superficiality—its mind is found wanting. When theory of mind confirms a user's suspicion that the device is either stupid (for example cannot remember what we were just talking about), or has only a limited grasp of obvious context, that user adopts dumbed-down behaviors in an attempt to match the perceived capabilities of the machine. When that action fails to correct problems, the result is user mistrust, abandonment, or anger.

Similarly, intelligent machine behaviors depend on the effectiveness of the machine's theory of the user's mind. This ties in with the memory problem. One of the key elements of an effective machine theory of a user's mind requires not only short-term memory of the immediate conversation, but a much longer-term maintenance of a personal and experiential context. Moreover, the machine must have a theory of its own mind—that is, a good understanding in the face of uncertainty of how it's doing and how the conversation is proceeding. The fundamental sophistication of a given pursuit is severely limited when the device cannot model the user, the history of the conversation, and itself.

Machine-Initiated Error Recovery

In a speech interface, errors manifest as illogical machine behaviors following user input. Users assume that the technology misrecognized, saying things like, "She didn't understand what I said." In other words, the ASR is not "accurate" enough. But thinking of all errors as "accuracy" flaws—misclassifications committed by the lowest-level acoustical attributes of the base technology—is a common mistake. Errors, in fact, occur throughout the layers of perception, cognition, and mutual negotiation that comprise an intelligent conversation. Furthermore, errors increase exponentially as applications become interactive. Anticipating them, detecting them and repairing them is not the exclusive job of the ASR vendor, it is the joint responsibility of the interdisciplinary team that releases a product.

Machine-initiated error recovery (of one sort or another) has been in place in IVR systems for some time. Designers are at least moderately comfortable with the concepts of word substitution, false acceptance/rejection and other perceptual errors committed by speech recognizers. Designers are also somewhat comfortable with

the notion that users need the power of review and confirmation, and that users sometimes change their mind about their desires. More subtle errors, however—either very low-level pre-perception timing problems caused by turn-taking, or very high-level cognitive errors caused by inappropriate theory of mind—slip through unnoticed, appearing instead as problems with ASR "accuracy."
The following machine-initiated recovery is from some early research that I and my colleagues did on number-capture dialogues as part of our GoodListener™ project many years ago (Balentine 1992).

Machine:	OK, what's the number?
User:	It's two-oh-one ... four-one-three ... six-two-five-nine.
Machine:	Two-oh-one ... four-one-three ... six-two ... **nine**? nine?
User:	No, **five**-nine.
Machine:	**Five**-nine. Got it.

Dialogue 2—Numeric entry and error repair

I choose the example for several reasons: 1) numeric capture is a recognition challenge that does not benefit from grammatical patterns or NL advances; 2) numeric capture continues to be an important task, still neglected by designers (who tend to prefer more interesting social dialogues); 3) except when numeric recognition is perfect, attempts at recovering numeric errors often bog down an otherwise effective ACS dialogue; and 4) basic low-level interactive units—including numeric capture (McInnes & Attwater 2004), yes-no questions (Balentine 2007), form-filling and list-selection—constitute interactive speech-behavior primitives that are reusable in a multiplicity of dialogues with all the usability advantages that implies. So the dialogue has relevance despite being both simple and old.

In the example, a common error is detected and repaired through a mixed-initiative user response. The repair is highlighted in bold. The interactive solution exploits common and basic attributes of all technologies, including n-best list, confidence, playback inflections, and a generalized turn-taking protocol.[4] This user, in subsequent interviews, did not recall any errors and reported that the recognizer was "very accurate."[5]

User-Initiated Error Recovery

User-initiated error recovery carries more difficult interactive baggage, including managing a global set of just-in-case user utterances in the grammar, using interruption times and garbage detection to trigger backward moves, and distinguishing between backing up to the immediately-preceding state (the most common user-initiated move) versus canceling the current action for a clean restart. Design of user-initiated recovery has been neglected due to the historical dominance of IVR applications: user learning does not play a significant role in IVR, and user-initiated error recovery entails behaviors that must be learned.

Interestingly, there are some excellent feedback loops in play between machine-initiated and user-initiated error repair. The machine can teach basic mixed-initiative behaviors to users during the normal interaction of a machine-initiated recovery—rewarding and reinforcing the behavior. Subsequent user experimentation then makes the error-repair behaviors intuitive and easy to discover.

Conclusion

Looking back at Figure 1, we can now ask, "What is the goal at (e)?" If that goal is a fully conscious, sapient and sentient entity that behaves in ways that approximate human conversational speech, then we must conclude that such a goal is quite distant—perhaps even further into the future than the distance already traveled to get to (a), (b), (c) and

---

[4] The example is extracted from a larger usability test that exhibited several classes of error, many of which were not recovered. Details on the algorithm, especially the handling of infinite regress, are beyond the scope of this article.

[5] A typical comment. Users do not remember errors that are quickly and gracefully recovered. Poorly recovered errors, on the other hand, generate user emotion and so dominate memory of the interaction.

(d). If we conclude that we are merely simulating such behaviors, then we're compelled to admit that even that goal remains elusive (albeit closer). If we conclude that we are designing user interfaces that serve users well enough that speech technology diffusion might approach the ubiquitous threshold, then that goal is within reach. But it won't arrive until we design interactive and robust multi-turn conversations that empower users to control progression toward complex goals.
Such an interactive multi-turn[6] conversation requires several very basic capabilities not currently exhibited by speech products:

- A stable turn-taking engine;
- Short-term memory of the conversation as it unfolds;
- A theory of mind model of both user and machine;
- Situated awareness;
- Knowledge about whether a goal is being achieved; and,
- Self-knowledge about how well each party is doing.

This is not an exhaustive list, but a pretty good starting point for achieving what end users imagine when they are told that a product is "intelligent" and "conversational." Of course, one-shot interactions sometimes pass the subjective intelligence test if the user has constructed a meaningful and information-rich utterance. But highly interactive, multi-turn dialogues are required if we are to achieve the ubiquitous speech interface.

Acknowledgements

Illustration in Figure 1 by Alexander T. Klein. Illustration in **Figure 2** based on Rogers, E. (1962) Diffusion of innovations. Free Press, London, NY, USA.

---

[6] A typical multi-turn interaction has a duty cycle of approximately 50%—that is, user and machine each talk roughly half the time.

References

Attwater, David, & Balentine, Bruce. (2009) "Turn-Taking Model," *U.S. patent #379607*, Washington, DC: U.S.

Attwater, David, & Balentine, Bruce. (2010) "Turn-Taking Confidence," *US Patent 7,809,569*. Washington, DC: U.S.

Balentine, Bruce. (1992) *GoodListener Cookbook*. Scott Instruments Corporation. Denton, TX

Balentine, Bruce. (2013) "'Super-Natural' Language Dialogues: In Search of Integration" in *Mobile Speech and Advanced Natural Language Solutions*. ed. By Neustein, Amy, Markowitz, Judith A., Springer Science+Business Media, New York.

Balentine, Bruce. (2007) *It's Better to Be a Good Machine Than a Bad Person*. Annapolis: ICMI Press.

Bass, Frank. (1969) "A New Product Growth for Model Consumer Durables," [sic] in *Management Science*. See http://www.bassbasement.org/BassModel/Default.aspx for a more modern discussion of the Bass Model.

Christian, Brian. (2011) *The Most Human Human: What Artificial Intelligence Teaches Us About Being Alive*. New York: Anchor Books.

Damasio, Antonio R. (1999). *The Feeling of What Happens: Body and Emotion in the Making of Consciousness*. Orlando: Harcourt.

McInnes, Fergus, & Attwater, David. (2004) "Turn-Taking and Grounding In Spoken Telephone Number Transfers," in *Speech Communication*. 43, 3, p. 205/223.

Pinker, Stephen. (1994). *The Language Instinct: How the Mind Creates Language*. New York: HarperCollins.

Rogers, Everett. (1962) *Diffusion of Innovations*, New York : The Free Press

Moore, Geoffrey A. (1991) *Crossing the Chasm*. New York: HarperCollins.

Moore , Roger K. (2015) "From Talking and Listening Robots to Intelligent Communicative Machines," in *Robots that Talk and Listen*, ed. Markowitz, Judith A., Walter de Gruyter. Berlin/Boston/Munich.

N. Ström & S. Seneff (2000): "Intelligent Barge-In in Conversational Systems," *Proc. ICSLP 2000*, Beijing, China.

Virtual Assistant Design Strategies for Behavior Modification and User Adoption

By Eduardo Olvera

The Age of Virtual Assistants

The idea of Virtual Assistants – digital services created to help satisfy a range of our needs – is not only fast becoming a reality but is also here to stay. These assistants are becoming our gateway to the internet, while leveraging artificial intelligence techniques to know more about us than we do ourselves.

In general, users of these assistants already come with certain expectations. Some examples are being accessible from any device while also being specialized and becoming better over time. Users also expect them to help in key areas such as personal daily needs, connections with people and places they know, and support for financial activities and financial management. They also assume these assistants are capable of integrating with crowd-sourced data to generate suggestions and recommendations. Finally, they expect them to be able to monitor environmental events related to things like weather or their health and well-being[7].

Aside from expectations, there are also nascent concerns around these assistants, particularly in areas such as privacy, confidence, trust, and transparency.

---

[7] Barth, J., Brauer, C., Goldsmiths College, University of London, Mindshare (2014, September 15). *What Can I Help You With?* Retrieved from https://www.mindshareworld.com/sites/default/files/What_Can_I_Help_You_With_Virtual_Assistant_Report_MindshareUK.pdf

Based on that, and compared to more traditional self-service solutions (over the phone, web, mobile, etc.), virtual assistants need to go beyond being able to just help users in real time or achieve a particular one-time goal.

The Ace Story

Back in 2014, a financial services company realized that some of their members were living paycheck to paycheck, and decided to do something about it. In particular, they wanted to help millennials become more financially savvy while helping them create better financial habits, starting with savings. Other studies (such as the one performed by Moody's Analytics) discovered that people younger than 35 (including millennials) were not only not saving money but also had a negative savings rate of 2%, meaning they were spending more money than they were making. In contrast, individuals between 35 and 44 years of age showed a positive savings rate of about 3%[8].

So the question was how to serve these members when faced with the challenge of not only not saving enough, but also very likely experiencing some significant life-changing events such as buying their first car, getting married, buying their first home or having a baby. As part of that, any solution needed to increase individual participation and engagement, advice members along the way, and change their savings habits to improve their financial health.

The answer? The creation of a standalone app, designed to reverse that trend. The Savings Coach app[9] was built from the ground up, included a personal financial assistant called "Ace", and leveraged

[8] Zumbrun, J. (2014). Younger Generation Faces a Savings Deficit. *The Wall Street Journal*. Retrieved from http://www.wsj.com/articles/savings-turn-negative-for-younger-generation-1415572405

[9] Savings Coach (2016). Savings Coach (Version 1.3) [Mobile application software]. Retrieved from http://itunes.apple.com

speech recognition, text-to-speech technologies and gamification techniques.

The rest of this chapter explores the design strategies and gamification techniques that were integrated into the savings virtual assistant, generated over $120,000 of collective savings during a four-month pilot, and achieved an average enrollment rate for recurring transfers of over 43% - more than twice than what was originally expected!

Please also keep in mind that for all these strategies and techniques, there's no reason why they need to be exclusive to virtual assistants, web or mobile, as they could easily be implemented today in any other self-service support channel such as an Interactive Voice Response system (IVR).

Virtual Assistant Design Strategies

Since the scope and user expectations regarding virtual assistants cover a wider range of needs, designers have to create new solutions that address at least three key features: education, guidance and tools.

1) Education

The educational aspects of any solution first focuses on helping users understand where they stand, and from there, allows them to learn and grow. This part is directly related to the user's end-to-end **experience** and has to take into consideration how the solution needs to evolve **over time**.

Many self-service solutions are currently built as "one-size-fits-all", assuming users don't have any previous knowledge of the system, and relying on instructions or error messages to provide additional information in cases where users struggle or proactively request more information. For example, IVRs tend to have a static structure that rarely adapts to different types of users, and even worse, are not designed to evolve over time as users gain more experience with the system.

In contrast, a well-designed system that takes the end-to-end experience and evolution into consideration, follows a scaffolding pattern that builds upon the previous step's knowledge and normally comprises four steps:

a) "Discovery" - this initial step involves elements that allow one-time users or visitors to quickly ramp-up and understand the basic features offered by a system or assistant. This part attempts to answer questions such as "What is this?", "It this right for me?", "What's the value it provides?", and "Will this help me get the answer or result I'm looking for?"
b) "Onboarding" - the second step of the journey focuses on users that come back but still have limited experience with the system (newbies). Think of it as the part that answers questions such as "How do I learn the ropes?", and "How do I start providing information and getting value?"
c) "Skill-building loop"- this is the third step and focuses on regular users. This functionality gets triggered by certain user actions or events. These same triggers allow users to respond and engage with a system, obtaining in return immediate feedback and information on their progress. Those actions then generate a call-to-action or provide an opportunity for those users to customize and personalize the experience, hence forming a habit-forming loop. This part addresses questions such as "What urge pulls me back?" and "What am I getting better at?"
d) "Mastery"- the final step in the journey, looks at the path for those enthusiastic users that become truly proficient in a system. It usually goes beyond the original scope of the system, and deals with more profound user questions such as "What

skills and knowledge have I mastered?" and "How can I leverage that?"

For example, when designing our virtual assistant "Ace", we defined a scaffolding pattern that started with educating users about their current financial situation (where they stood, how much money they had, how much they were spending, on what and how fast). From there a natural progression would be educating them about savings and how to get started (including the notion of "pay oneself first" and creating an emergency fund). Once spending and savings are under control, you start looking at ideas such as good debt and how to take advantage of it, to eventually get to a point where advanced users can be educated on investments, having money work hard for them, understanding diversification and other financial instruments like insurance.

2) Guidance

Closely linked to education, any solution needs to also take into account the idea of guidance or advice. As discussed earlier, users expect virtual assistants to become smarter over time, and to be able to provide relevant and personalized suggestions and recommendations. A critical part of that guidance is that the system should explain things in terms a user can understand, via simple concepts and examples, while also avoiding the use of jargon or complex terminology.

This idea of guidance can be naturally represented by the idea of an "assistant" or a "coach". There are many applications and systems out in the market that depict these guiding traits as well as the notion of an advisor (see Figure 1), for example:

- Mint[10] – Financial assistant that focuses on money management, budgets and personal finance

[10] Mint.com. (2016). Mint (Version 4.10.0) [Mobile application software]. Retrieved

- Acorns[11] – Financial advisor that allows users to invest spare change into a diversified portfolio
- Toshl Finance[12] – Personal finance manager focused on savings, budgets and expense tracking
- Fitocracy[13] – Personal fitness coach with workout exercise logging for weight loss
- MapMyRun[14] – Running coach with GPS and workout tracking with calorie counting
- MyQuit Coach[15] – Health advisor that supports you in quitting smoking

Figure 1

from http://itunes.apple.com
[11] Acorns. (2016). Acorns (Version 1.5.8) [Mobile application software]. Retrieved from http://itunes.apple.com
[12] 3fs. (2014). Toshl Finance (Version 1.8.7) [Mobile application software]. Retrieved from http://itunes.apple.com
[13] Fitocracy, Inc. (2015). Fitocracy (Version 3.5) [Mobile application software]. Retrieved from http://itunes.apple.com
[14] Under Armour. (2016). Map My Run (Version 16.7.3) [Mobile application software]. Retrieved from http://itunes.apple.com
[15] Demand Media, Inc. (2013). LIVESTRONG MyQuit Coach (Version 2.1.1) [Mobile application software]. Retrieved from http://itunes.apple.com

For "Ace", we decided to define it as a "Fitness tracker for your finances", with its value coming from the fact that people who track their finances have better financial health. From this tracker, the idea evolved until we decided to have a "financial coach", while its persona evolved over time and as a result of multiple iterations and numerous usability tests (see Figure 2).

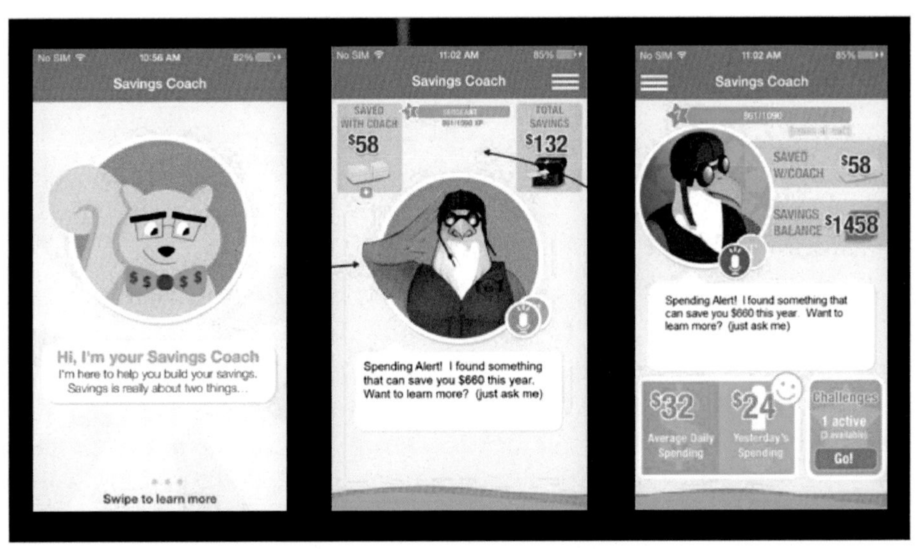

Figure 2

One thing to note is that once you define a personality for your system (or assistant in this case), it gives you a lot of freedom in terms of design of the UX. As you can see in some of the verbiage above, the persona helped guide the dialog and allowed for some creative (and fun) interactions along the way, such as the idea of "squirrel some money into savings" or "don't go nuts with your spending" during the first squirrel iteration.

3) Tools

This is the support and reinforcement component of your solution. The idea is that once your users know where they are and what they can do, and the system has a clear and extensible path, then the third supporting block is the tools and techniques that make it easy for users to set things up, to automate repeating process, to turn complex interactions into no-brainers or one-step transactions.

These tools are also closely related to the skill-building loop defined earlier, where tools like personalized data, proactive notifications and calls to action, yield better user engagement, allow users to be more invested in the system, and promote positive habit formation.

For "Ace", one of the key tools we leveraged as part of our design was gamification, which is the topic we'll discuss next.

Gamification Elements

Once visual elements, design components and overall scaffolding is defined, the next step is to think about design strategies. For our coach "Ace", we had a very clear goal (get members to start saving) which was to be accomplished via the setup of recurring transfers. Since this was something members weren't used to, that meant we needed to find strategies that facilitated behavior changes and user adoption.

One of those strategies is gamification. This term sometimes causes confusion amongst designers as their definitions don't always make a distinction between gamification elements, and gamification techniques. An easy way to think about this is that elements are the equivalent of a "toolbox", which includes things like leaderboards, points, quests, avatars, social graphs, levels, badges and rewards. This is where most people stop and assume that's all there is about Gamification – often referred as the "PBL Triad"[16] for Points, Badges and Levels – which makes them believe that by simply adding these elements, any system can be "Gamified". Also, gamification is not

---

[16] Da Silva, D. G. (2015). Points, Badges, and Leaderboards [Web article]. Retrieved from http://www.agilegamification.org/gamification/points-badges-leaderboards/

about making everything a game, the use of games in business, or simulations.

In our case, we're looking at the more encompassing definition of gamification: "The use of game elements and game-design techniques in non-game contexts."[17] The techniques refer to the different ways in which you can approach it, such as motivation, psychology, and behaviorism, and the non-game contexts include areas related to positive social impact, business advances, or personal growth.

To better understand this concept, let's think about games for a second. Why do people play games? What can we learn from games?

One of the most powerful elements of games is that they give players a sense of ***autonomy***, defined as a set of meaningful choices that yield results that players can actually see. Also, games normally have rules, but even then, there's a sense of freedom in a game as it gives you free motion within that set of constraints. Combining these concepts and the idea of thinking about our users (customers, employees, participants, etc.) as "players", we can take a slightly different approach and follow a "***player-centered design***", where our system can give users a certain sense of autonomy as well as a certain degree of freedom to decide what it is they want to do next.

Furthermore, we can think about the notion of a "***player journey***" – that conceptual path players follow through a game – from onboarding to mastery that gives a constant sense of progression and works hard to achieve an ***integrated experience*** – color, visual design, dialog and sound effects, all working together to create a deeper and richer experience. Strive also for a sense of ***balance***, so that your system's activities are not too hard or too easy (which can make an experience boring very quickly), and that you don't limit the number of choices too much, or give them so many things to choose

---

[17] Werbach, K. (2012). *For The Win.* Philadelphia, PA: Wharton Digital Press.

from that they may get paralysis. Finally, you add ***emotional*** components that make an experience more engaging and fun!

Yes, just because we have business requirements and technical limitations, doesn't mean a solution cannot be fun. Fun can be designed. And people have very different concepts of fun.

Let's do a little experiment. I'm going to ask you to please play along. First of all I want you to close your eyes for a minute (OK, please read the instructions first and then close your eyes), and think about the last time you really felt you had fun, when you truly enjoyed doing something. Think about an experience where you really savored the moment, where time just flew by... Now I want you to try to remember how you felt afterward. Ready? Go!

All done? Good times, right? Perfect. Now here is a list of reasons and I want you to try to map that experience to one of the reasons for why you performed that activity in the first place:

1. Did you do it because it solved a problem?
2. Was it because it was something you wanted to do?
3. Was it something that allowed you to interact with other people and socialize with them?
4. Did you do it because it allowed you to create something new?
5. Was it something meaningful to you?
6. Did it give you a sense of meaning or allow you to help others?
7. Did it resulted in a tangible benefit like money or something else of value to you?
8. Was it something that gave you a sense of control?

There are no right or wrong answers here, and maybe you picked more than one. All these seemingly random reasons actually relate to the top three factors that motivate someone to do something and consider something to be "fun".

If you picked reasons 1, 4 or 7, that means that one of your main internal drivers is ***Competence***. Competence means that you enjoyed the experience because it gave you a sense of ability, a sense of accomplishment, that feeling you get when you achieve something or earn something as a result of your work.

In contrast, if you selected reasons 2, 5 or 8, you tend to be driven by a sense of ***Autonomy***. Autonomy means that you like having a good sense of control, enjoy the sense of freedom and having the choice to do things that are meaningful to you.

Finally, if you ended up with reasons 3 or 6, that means ***Relatedness*** is your motivating factor. Relatedness is that sense of purpose, of community, of having a connection with something beyond yourself, and having the ability to have a social interaction with others like you.

All these factors are what is commonly referred to in Psychology as **intrinsic motivators**. These are the things that move people to do things from within. In contrast, there are other factors called **extrinsic motivators** such as physical rewards, money, scores or grades that come from the outside.

The reason we draw this distinction is because in order to modify behavior and promote good habits, intrinsic motivators have a much stronger pull on people. Therefore, in order for a solution to yield positive results and increase its "sticky" factor, you should pick more than one of these motivators and integrate them into your solution, so as to make it more appealing to all these different types of "players".

Putting it all together – Gamification Techniques in Action

Based on the objective of our project, the educational aspects we wanted to cover, the gamification elements we considered relevant to provide guidance and allow users achieve their goals in a fun and engaging way, here's a behind-the-scenes look at how we implemented and integrated five specific gamification techniques:

1) Onboarding

As discussed in the Education section of VA strategies, for the first iteration we focused on the Discovery and Onboarding aspect of our solution. To begin with, we created a sequence where the VA would introduce itself and set up the right expectation from the very beginning, letting users know about what the solution was about and why they should care. Things like "I'm going to be helping you with your savings" and "you are going to be able to earn rewards along the way", create a first-time-user experience that educates users the first time they open the application, which then gets disabled for subsequent launches of the application (although we added functionality under the application settings that allowed users to re-enable it).

The second part of the onboarding process included a sequence of steps that focused on helping you set up an automatic savings plan. To simplify that experience, the system uses "smart defaults" to prefill as much information as possible (as opposed to having users fill out a lengthy form) based on known user spending habits, past user behavior, and account balances which personalize the interaction. With that information the system would make a recommendation and simply ask users "Will that work for you?".

Furthermore, we designed this initial interaction to be completely foolproof. For example, there was no speech component to it yet. We wanted to introduce the concept, allow users to understand the basics of the application, introduce them to our VA, and get them signed up for recurring transfers with a single tap. Since it only involved simple gestures (like taps) and all information was prepopulated from user's accounts, we made sure users wouldn't get stuck or fail on that critical initial interaction, which also increased the user's trust in the VA and the system in general.

2) Reward Schedule

As a way to entice users to accomplish certain goals, we evaluated the use of rewards. From a gamification perspective, there are different types of rewards that can be applied under different conditions and contexts. Since each come with their pros and cons, we decided to implement a mix to maximize their impact. In our case we integrated four types of rewards:

a) Fixed Ratio Rewards – the concept of fixed ratio is that users will get rewarded in a consistent way every time an action takes place. In our case, since one of our goals was to get users to save money, we decided that for every dollar someone saved, we were going to give them one point. These types of elements become things that users come to expect and have no limit to them. Unfortunately, even though they add consistency to a solution, they tend to become boring very quickly as users expect them to happen no matter what.
b) Variable Ratio Rewards – another type of reward that looks at giving users something after a certain number of times that an action takes place. In our case, we added specific rewards for things such as setting up your first automatic transfer or completing six challenges. These tend to be a little bit more complex than fix ratio rewards, but because of their variable ratio, they tend to be more surprising, and hence are more interesting to users.
c) Fixed Interval – these types of rewards are similar to ratio rewards, but instead of looking at the number of times something happens, they take into consideration the timing of those events, or the interval between them. In this case, we looked at opportunities to reward users after specific time intervals, such as logging in five days in a row (every 5 days), and even having a special one if you log in the day of your

birthday (every 365 days), which give you an unexpected benefit on a day that tends to be special for people, hence connecting with the emotional side of an experience.

d) Variable Interval – finally, these are the best types of rewards as they do not have any type of schedule and yet have the most psychological value because our brains love surprises! In our case we selected a few things that we didn't really advertise that much, but ended up wowing users whenever they came across them. For example, we reward users after interacting with the VA for 10 times by giving them a badge, or when they complete three challenges in a row.

3) Medals

In gamification terms, this was our equivalent to traditional "badges". In our case we made sure each medal was linked to specific actions. We linked them to triggers and behaviors we wanted our users to perform, as a way to reward their behaviors and increase the adoption rate. For example, setting up automatic transfers was a key goal of our solution, so we decided to award users a medal the first time they set one up. Aside from giving users medals, we had to think about the best way to educate users about all the ones they had not earned yet. From usability testing, we found that showing them in the screen (with a different visual treatment to distinguish them from those already earned) yielded two big benefits. One, it served as a teaser as it appeals to our innate sense of completion – people crave to complete sets and finalize collections they have already started. Second, it allowed us to educate people as tapping into a "locked" medal gave users information about what they needed to do in order to earn that medal.

4) Challenges

As we covered in the topic of intrinsic motivators, we thought about giving users a sense of control and autonomy over our system. For that we implemented challenges. And we wanted to start very small and simple. In our case, we included a basic set of challenges in order to collect user feedback and usage data. We wanted to make sure challenges were meaningful to people, and initial data revealed these items were things users could relate to – substituting coffee for water, pay yourself first, skipping pizza – with different frequencies and durations. Challenges were small enough for people to be able to achieve them, but meaningful enough to start forming some healthy habits.

Furthermore, having a set granted users a sense of autonomy and freedom as they had full control over which challenges to take and how often to take them. Finally, we also focused on what would happen when things didn't work out, such as when a challenge deadline expired or the user was unable to complete it in full. In those cases we added a mechanism to reset a challenge. And because the VA was designed with a "coaching" sense, we wanted to emphasize the positive aspects of the experience by creating interactions that acknowledged what they accomplished so far (making them feel good about themselves), and steer them towards completing it in full (e.g. "You only have two more to go!"), giving them a chance to get back on track and trying it again.

5) Notifications

Finally, we looked at ways of creating an experience that was immersive and proactive. For that we leveraged native features of mobile devices and created the equivalent of "proactive notifications" from traditional IVR systems. In our case, we created a list of items that we wanted to remind users about, some related to their challenges, some to their goals and points, and determined the best way and time to inform users about those activities. We also added prioritization to that list (keeping actionable items and things users could do at the very top) and included personalized elements as part of those notifications (name, outstanding number of actions, needed, etc.).

In terms of timing, we were also very careful about keeping things moving (so users wouldn't get bored or lose interest) while avoiding bombarding them with messages every time they pulled out their phone. For that we added rules to avoid duplications or having the same things pop up too often.

And in terms of proactivity, we made sure to wrap up every notification with an actionable bit, such as "Ask Me" or "Tell Me" that focused on getting people engaged with the solution. Once a user receives a message on their device, they can act upon it, which then takes them directly into the application to interact with the VA coach. At that point the assistant will present the same message, so that users can continue the conversation within the system, hence minimizing interaction barriers and giving users the sense that the coach was active and looking out for them all the time.

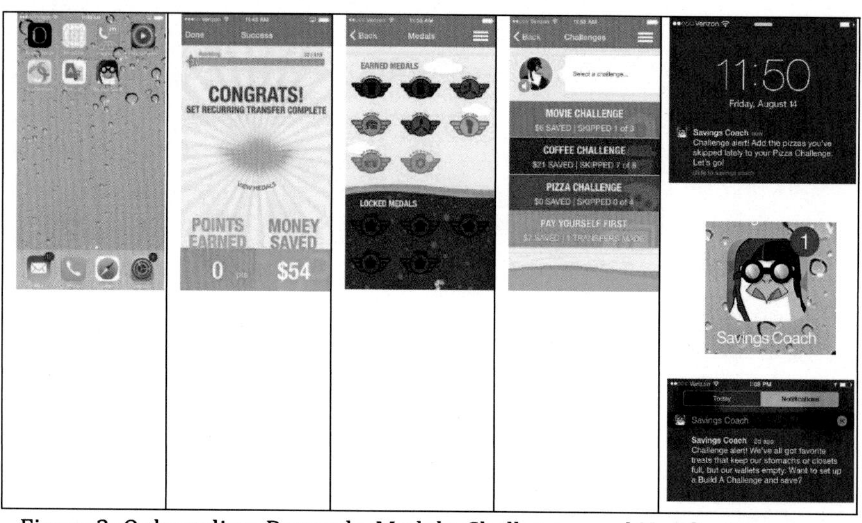

Figure 3: Onboarding, Rewards, Medals, Challenges and Notifications

Final thoughts

As we've seen, there are design strategies that can be applied to any system, including virtual assistants that can help with behavior modification and user adoption. In particular, thinking about your users as players, finding out what motivates them, and leveraging gamification elements and techniques, have shown fantastic results when properly designed and integrated into a solution.

In our case, ongoing usability testing uncovered some interesting findings that can be applied to any type of system:

- Users really appreciate systems that provide useful information, particularly around things they didn't know and that have impact on their personal life.
- Their expectations go beyond that, particularly when the notion of a "coach" or "assistant" is introduced. Users expected the VA to do the analysis for them, find really good solutions in their behalf and help them save more, faster.
- Furthermore, they also expected the VA to focus on "top of mind" features, meaning that they would like for the application to know about what the user is doing (e.g. walking into a coffee shop), intercept them in the moment, and proactively probe them (e.g. "Hey, do you want to take on the coffee challenge today?"), giving them an opportunity to change their behavior and reinforce a new habit.
- User quote: "You made a game out of saving, genius! I don't put anything into savings now, so you guys are helping me out big time! Thanks."

And the results? As part of the tuning and evaluation process, the application ran during a four-month pilot with roughly 800 participating members, ages 18-24 (our millennial target audience). Together, they had a collective savings total of almost $120,000 during that period[18]. Furthermore, the enrollment rate was very surprising. The original goal was to target a 10%-20% rate, but during

---

[18] Wisniewski, M. (2015). Launch of New App to Help Millennials Save. *American Banker*. Retrieved from http://goo.gl/JdqPMo

the pilot period the actual number was a steady 43%, which stayed strong even after the application went live.

Think about the systems you're designing. Is there anything you can do to reward their behavior? Can you make them want to self-serve? Can you find an emotional connection with them and tap into what moves them internally to do something? Can you make it fun?

Your turn.

Just Like Talking to a Person? Next Steps in Natural Language Understanding

By Deborah Dahl

Spoken interaction with technology, whether with our mobile devices, web applications, or physical devices in the Internet of Things, has come incredibly far in the last few years. After decades of promises that spoken language technology will make it possible to interact with a computer "just like talking to another person", we are finally starting to get there. There are now a number of general-purpose assistants -- Apple Siri, Amazon Alexa, Soundify Hound, Microsoft Cortana and Google Assistant, for example. These systems support spoken natural language queries on such topics as times and dates, weather, news, sports scores, and many other kinds of everyday knowledge. The quick, voice-enabled access that these systems provide to everyday information is extremely convenient. In addition, there are also quite a few freely available tools for developers to build their own personal assistant applications. These tools include Facebook's wit.ai, Google's DialogFlow (formerly api.ai), Amazon's Alexa Skills Kit, Nuance Mix, and Microsoft LUIS.

As impressive as these systems are, their natural language understanding capabilities are still very limited compared to those of people. Is talking to them "just like talking to a person"? Sure, if the person is four years old. Talking to the Alexa system, for example, has been described as "like talking to a toddler"[19]. Of course, most four-year-olds do have pretty good language abilities, but they're still

[19] T. Baker, "Amazon Echo - Life with Alexa," Available: http://tobybakersparklingstories.blogspot.com/2015/12/amazon-echo-life-with-alexa.html

learning, and current natural language understanding systems are a lot like them in that way.

Just to get a little better handle on what today's systems can do, I performed an informal test of the natural language understanding abilities of some personal assistant applications and presented it at Mobile Voice 2016[20]. My test included about 150 queries in a variety of domains, including:

- Time and date: "What time is it?"
- Everyday knowledge: "How tall is Barack Obama?"
- Weather: "What's the weather forecast?"
- Local business search: "Where is the nearest Chinese restaurant"?
- Reminders: "Remind me in 5 minutes to ..."
- Arithmetic: "What is the square root of 768?"
- Timers: "Set a timer for 5 minutes"
- Out of domain – questions that the system could not possibly know the answer to, such as "What color is my shirt" Systems should be able to respond that they don't know when they get a question like this.

Although all of the questions were spoken to the virtual assistants rather than typed, I didn't want to test speech recognition, just natural language understanding. I only looked at utterances where the speech recognition worked perfectly.

In addition to different domains, my test also included questions differing in syntactic, semantic and pragmatic complexity.

---

[20] D. Dahl, "LUI 2.0: Next Steps in the Language User Interface," presented at the Mobile Voice Conference, San Jose, California, USA, April 12, 2016.

"Complexity" can be somewhat subjective, but I considered that the following aspects of an utterance would make it "complex":

- Omitted information: Pronouns (anaphora) and ellipsis: "Who was the tallest American president?" followed by "Who was the second tallest?"
- Comparisons: "Which has more calories, broccoli or an apple?"
- Beyond the here and now: Time expressions, possibility, probability, contingency, negatives: "Are there any dwarf planets other than Pluto?"
- Indirect requests, for example utterances including implicatures: "It's time for Japanese food", has an implicature, based on the Maxim of Relevance, of "Where is the nearest Japanese restaurant?"

While basic queries in all of the domains were handled well by most of the systems, the complex queries were very difficult, and not handled at all well. And, it's worth pointing out that, while these queries were in general too complex for most of the systems to handle, they aren't very difficult for people. The complex queries are very natural. Of course, getting answers to the basic queries is unquestionably very useful. People are also generally good at rephrasing their questions when the person they're talking with doesn't understand. But constantly repeating yourself is tedious, and I believe these assistants could be even more useful if they were better at handling queries that are even a little bit more complex.

Let's take a closer look at one specific category of questions – those that go beyond the here and now, such as "what if" questions. Many domains lend themselves to a slot-filling type of interaction, and local business search is a good example. Slots generally correspond to features that the user is interested in. For example, in restaurant search, the user might be interested in such features as the type of cuisine, the location of the restaurant, the number of stars, and cost. Wouldn't it be useful if we could ask about hypothetical situations with slots filled contingently, depending on other slot values? To be concrete, what if we could say things like "I'm looking for a Thai restaurant that's close by, but if it's more than 3 miles away, see if there's a closer Vietnamese one." So, filling the "cuisine" slot in a restaurant search application with "Thai" or "Vietnamese" is contingent on the distance of the nearest Thai restaurant. This is something that most people could understand very easily. The only way get this information from today's systems is to break it down into a search for "Where is the nearest Thai restaurant?", and then if the nearest one turns out to be too far away, start another search for "Where is the nearest Vietnamese restaurant?"

This strategy of making the user break down an utterance into simpler components and then mentally synthesizing the results amounts to making the user micromanage the conversation and puts an extra burden on the user. This burden results from both the need to come up with the unnatural simplified utterances as well as the need to mentally put together the results to synthesize the final answer. Not to mention that it also takes more time. Since simple slot filling is well within the capabilities of today's natural language systems, it should not be too big a step to enable systems to maintain a couple of alternative sets of slots and represent the user's criteria for selecting the best alternative.

One problem with trying to build systems that can handle more complex queries is that the complex queries are much more varied than the simpler ones, and each different type of complexity occurs less often. So dealing with them is more work with a smaller payoff than handling the simple queries. But I believe we need to start looking at these if we want talking with virtual assistants to really be like talking to a person.

The natural language understanding capabilities of today's virtual assistants and toolkits are truly amazing, but they are far from "just like talking to a person". Much more could be done to bring their capabilities closer to the level of what we can do. Now that systems can handle the simpler queries well, developers should start looking at the long tail of more complex, varied and natural possibilities.

Adding Voice Interaction to Hardware Products

By Leor Grebler

Over the past four years, an increasing number of consumer hardware products have started to come to market that have voice interaction built into them. However, 2016 will prove to have been a pinnacle year for voice interactive products coming to market: Super Bowl ads for the Echo, Google Home's release, and dozens of products hitting the market from major brands with voice interaction built into them.

It's not surprising that many devices makers are looking at their competitors that are engaged in voice interaction and are evaluating whether they too should pursue this path. When looking at their options, there are several factors that they need to take into account when making their decision. Some items that they need to consider are: what type of interaction do they want their users to have with their devices; what are the bill of material (BOM) considerations; what services and brand experiences are they considering; and what expandability would they like to implement for future enhancements.

Types of devices

The biggest influencer on the type of voice interactions that needs to be built into a device is the device type. For the purposes of analysis, we can look at four types:

Utility devices

These devices are products that are typically meant to perform some type of physical action or complete a chore. Laundry machines, dishwashers, toasters, microwaves, and vacuum cleaners fall into this category. People who use them typically acquire them to complete a task rather than use them for any prolonged enjoyment.

On a personal and home use basis, the argument for adding voice is to speed interaction with the device and make it easier to use so that the device can reach a larger market. For professional applications, the argument for voice needs to also encompass an ROI measurement. Can using voice create more worker efficiency?

Entertainment devices

These devices are used for enjoyment and recreation and include music players and televisions. They are used primarily to consume content or media. Also included in this category are accessory devices such as TV boxes.

Considerations for adding voice to these devices are whether doing so enhances the enjoyment and pleasure of the end user, differentiates the product, or potentially creates an additional stream of revenue for the manufacturer or brand.

Office devices

Office devices can take the form of conference calling equipment, projectors, landline phones, photocopiers, or other utility devices that are specifically used in the conducting of business. There are other devices such as thermostats or lighting controls that could potentially fall within this category.

Brands and manufacturers of these devices will likely need to look at whether offering voice control with these products provides a time savings advantage or reduces distractions. The argument needs to be made that this feature will increase office productivity.

Computing devices

Mobile devices and computers (desktops, laptops, tablets and their hybrids) already have extensive integration with voice interaction services. It's likely that in the coming year, more voice services will be available on computers and mobile devices will gain far field voice capabilities.

Considerations

Types of voice interactions

Hardware makers can consider five different types of voice interaction. The first is push-to-talk, where the device is woken up by the user performing a physical action and then speaking to the device either during the physical action or immediately afterwards. The second type is the command-based interaction, where the device is constantly listening for a specific command for the user and is continuously polling the microphone signal for this command. A hybrid of this type of interaction is with a primary wake up word followed by a secondary command. The third type is the free form interaction. This is preceded by a wake up word following which the user can speak using "natural language" - a request that doesn't require the memorization of a specific word or word order. The fourth type of interaction is one in which the user is prompted for a response. The response can be in the form of natural language or fixed command. The fifth type is remote voice interaction, where the user interacts with the device through an app on a computing device.

BOM and implementation considerations

The effect of the type of voice interaction on BOM requirement is considerable. For example, assuming that a device is already Internet-connected, adding voice interaction through a remote device, such as through an app on a phone, has no impact on BOM cost. This is an easier first step to implement voice interaction and can be rolled out to legacy WiFi-connected products that have already been shipped to customers. Applications could be playing music, turning the device on an off, controlling volume, among other actuations.

For devices that are command-based only, including push-to-talk followed by commands, the requirements for voice interaction can be fairly light. There will need to be some type of processor, whether a microcontroller or application processor, that will be able to run the phrase / word spotting or speech triggering software. Typically, software for phrases spotting can be designed to be so lightweight that it can run alongside existing applications on the same processor.

The amount of memory and processing required to run the software is proportional to the number of trigger words and commands as well as how specialized is the application. Local automatic speech recognition (ASR) tools, such as Pocket Sphinx or Nuance Vocon, require tuning of the ASR engine and typically have a sample period in which they look for a particular phrase. Since they are running continuous speech recognition, the memory and processing requirements for these types of implementations can be fairly large.

However, specialized local phrase spotting software from companies like Sensory, Malaspina, and Rubidium typically work after training against many recorded samples. The result is that these implementations can run on very low power chips. This makes them especially well-suited for battery-powered devices.

For command-based devices, there's a need for at least one microphone to be embedded on the device. Likewise, if there is any acknowledgement of the command, there might be requirements for LEDs or a speaker to play out a sound file, recorded speech, or a text to speech file. Depending on how well the command words are trained and the number of commands, it is possible to get a high success rate even at far field distances (two meters or more) from the device.

For devices that use push-to-talk or have a wake up words but are designed to be used in close proximity to the speaker, followed by a natural language command, it's possible to use a single microphone without any or with only very little audio processing. If the device, (such as a voice interactive headset, or a speaker with a push-to-talk button) are being used at less than arm's length from the user, the signal to noise ratio may likely be sufficiently high to allow for the automatic speech recognition (ASR) API to pick up and properly interpret the user's voice commands most of the time. There may still be sensitivity to background noise that could require the user to speak louder or more closely to the microphone.

As soon as the device's planned usage goes beyond arm's length from the speaker and the interaction goes from command to natural language, the hardware needs to condition the audio before sending it to the ASR API. This means the implementation of some type of digital signal processing algorithm and likely the addition of at least one more microphone. The reason for the additional microphone or microphones is to be able to implement algorithms such as blind source separation or beam forming.

Depending on the profile of the environment in which the device will be placed (kitchen, conference room, or next to the TV), hardware makers may need to increase the number of microphones from two to up to seven. This is to compensate for reverberation or loud noises that can affect the quality of ASR results.

Services and brand experience

Another major factor that hardware makers and brands need to consider before adding voice is the brand experience for end users. The voice speaking to the user from the device acts like the brand's mascot. The qualities of the interaction are reflected directly on those of the brand. If the answers are incorrect from a device or the user has issues in being understood, these qualities will be associated directly with the device. As an example, one could argue that Siri's initial release had a negative impact on Apple's brand.

Beyond working well, which is a given, voice interaction should have some connotation of the brand. However, as of writing, there are only two implementable options currently available to hardware makers - build end-to-end voice interaction using components from one or various service providers (trigger, speech recognition, natural language understanding, integration, and text to speech) or use Alexa Voice Service (AVS). The latter when used in far field hands free capacity requires "Alexa" as the wake up word. As well, device makers using AVS do not have control over the interaction beyond integrations to control the device. Audio is streamed to AVS and then the response is streamed back to be played out of the device's speakers.

It is likely that other players will step in with end-to-end and potentially customizable solutions. However, today, device makers need to consider whether the added benefit of having AVS as a feature of their product offsets the branding issue of users referring to their product as "Alexa".

Expandability and future-proofing

Technology in the voice arena is changing at an increasing rate. In 2016, there have been major announcements about new voice technologies nearly every month. This has caused some device makers to be hesitant in moving forward with adding voice control, hoping that the market will settle on a particular solution. The downside to this strategy is missing out on adding innovative and differentiating features to products and potentially giving up market share to competitors who are willing to take on the risk.

One way to offset this risk is to build in expandability into the product. While there are pressures to use only the bare minimum processors and memory to gain functionality, it's better that device makers build in some room as well as infrastructure to update the device with newer software as it's made available. This might mean ensuring that devices can check and receive updates and that API client version changes can be pushed down to the device.

Opportunity

While many device makers are currently looking to add voice, there is only a small minority actually making the leap to adding this feature to their products. Those who are early are likely to build a following and gain support of the major players in the space. It's also possible that they'll become the leaders in an emerging category of products. At the very least, those companies that have thought about adding voice control would put themselves at an advantage to begin to develop proofs of concepts and prototypes of their products that include voice.

Convenience + Security = Trust: Do you trust your Intelligent Assistant?

By Maria Aretoulaki

Each of **us owns more than a single smart device** at home, work, and on the road: apart from our inseparable companion, the smartphone, we may now have a connected smart watch, an Amazon Echo or a Google Home, as well as the standard tablet and laptop. They all need to somehow work together or compete for our attention and for ultimate control or at least collaboration on a task (e.g. which device should remind us of an upcoming event?). This increases the **need to sync and share data and tasks** across devices, something that has become simple and straightforward through the use of a single identifier, usually an email address and the associated account (e.g. Google, Apple, or Amazon). By logging in with the same username across devices, syncing is automatic and seamless, but **all of our personal, private and confidential and often sensitive data and services have also been migrated to the cloud** (the Google, Apple or Amazon servers that also happen to sit in the US). This includes our online profiles (Google, Facebook, LinkedIn, Twitter), our calendars with all the events we have attended and intend to attend, all the trips we have made and are due to make, all the people we have met and will be meeting, our location history and all associated digital histories (browser search terms, YouTube views, itineraries we take). All this together can say a lot about our personal preferences and behaviours, past and present, and can even predict with some accuracy our future preferences and behaviours too.

Personal Data

Despite the disparity of the types of personal data used and shared across devices, platforms, cloud services, applications and social media, it can all be unified, integrated, mined and harvested to **form a**

single background context completely personal to us. Technologies such as Information Extraction and Data Mining, Analogical Reasoning and Deep Learning, Ontology building, semantic and Knowledge Representation can be employed to generate customised interpretations against this context and draw relevant inferences in order to identify and prioritise courses of action and determine the optimal next task to carry out. Ubiquitous Voice Assistants on our smartphones, in our homes and cars - from the pioneering Apple Siri and Google Now, to the Google Assistant and Google Home, Amazon Alexa and Echo, Microsoft Cortana, Samsung Bixby and the Apple HomePod- can exploit this personalised context to **render speech recognition more accurate and robust, and natural language understanding more intelligent and individual to us.** If the Assistant knows your specific location in the house, it can limit the interpretation of a voice command such as "*Switch the light on*" to mean just the lights in the room you are sitting in, rather than assume it could mean any of the lights all over the house and having to ask for a disambiguation. If it knows from an email confirmation that you have a flight coming up to Athens, Georgia, it can limit the hotel search to just that city without having to ask you whether you mean Athens, Greece. Thus, Voice Assistants can naturally unify and give immediate, relevant and predictive access to and control of diverse devices, services, apps and data and provide personalized information, feedback and warnings and even predict what you want to do next. This makes them attractive and, soon, indispensable, given the amount of data each of us generates and needs to access and the number of cloud services we use daily.

Unwanted Data Mining

It doesn't stop there, however. **Organisations aiming for our business or loyalty** can also mine this disparate personal data to discover patterns, correlations and interdependencies to interpret, infer and even predict our intent. This way they can target specific needs and preferences we seem to have and successfully promote and ultimately sell certain products and services over others. **More**

sinisterly, however, governments trying to check up on us, and potentially control us, also have access to and can harvest this personal data. The most commonly stated rationale behind this nowadays is the prevention of terrorism and other crimes. Nevertheless, the accuracy of these interpretations and predictions actually depends on both the sophistication of the Data Mining and Machine Learning software used and on the quality and quantity of the data mined itself. It is easy to arrive to the wrong conclusions: Does searching for a controversial topic make you a suspect of planning to engage in the associated – potentially illegal – activity? And how long will that search term remain in your digital history? How will it affect the assessment of your character, financial status and health in the future by governments, credit reference agencies, potential employers and insurance companies? **Does sharing your personal data in the cloud make you more prone to such misinterpretations of your intentions and misjudgements of your character?** These are all issues that need to be considered very carefully and urgently, as ubiquitous Always-On Assistants become more prevalent and permanently listen to and record everything we say and do.

Life Sharing vs. Privacy

With the dominance of Facebook and similar **"life sharing" platforms** (e.g. Twitter, Instagram), which encourage sharing with the world our successes and failures, relationships, family events and illnesses, political and religious opinions, we are already accustomed to sharing private and often sensitive information (such as photos showing your home address). This has made it natural to share information that in previous decades no one would even dream of sharing or would be too embarrassed or wary to. As more data is accumulated on and by each and every one of us, technologies are continually being developed to exploit and make sense of as well as monetise this data, giving rise to very personalised and contextually adept Intelligent Assistants and chatbots for personal and work organisation, health and illness management, relationship management and other innovative products and services. Big Data is crucial to generate and monetise such intelligence. Nonetheless, this "**sharing culture**" goes hand-in-hand with growing concerns about data security and privacy, even if we don't actively do much to promote either. And the pessimists and conspiracy theorists were recently proven right, to their own dismay[21]. Thus, a major driver that will determine the success and level of adoption of Ubiquitous Voice Assistants and similar services in the future is the **degree of trust they inspire in their users**. Whoever can guarantee and ensure the safeguarding of data ownership, ultimate long-term control and security against the prying eyes of governments, dictatorships or your average scammer will ultimately gain our loyalty - and our money.

Figure 1: This is your privacy online[22]

Data Control

Even if you don't share any private data on social platforms, **data backups** are important too. Think of the times you had your phone lost or broken, but still found – with great relief - all of your photos

---

[21] https://www.theguardian.com/technology/2018/mar/24/facebook-week-of-shame-data-breach-observer-revelations-zuckerberg-silence

[22] http://thedailydose.com/comic/this-is-your-privacy-online/ - August 2011

and videos from the past 2 years waiting for you in the Cloud! Still, data permanence may be undesirable or even dangerous in some cases. Even if you have **“nothing to hide” from your prying government**, there is no guarantee that your personal and sensitive data will never land in the wrong hands. It could be accessed by playful hackers, ruthless criminals or nowadays more often than not greedy marketers, who in the best case just want your money, in the worst may want to steal your identity, or even endanger your family’s safety. Researchers are now experimenting with remotely hijacking the electronics of (self-driving) cars to prove they suffer from serious security holes or embedding voice commands you can’t hear in your Intelligent Speaker[23]. It is no longer unthinkable that someone can plant voice commands pretending to be you suddenly taking over control of your car or home security system! A new era in sophisticated car theft and home burglaries![24] To take the argument further, if a malicious third party manages to steal your identity, there is nothing stopping them from posting damaging content on the internet on your behalf, which may irreparably harm your reputation. For this reason, **the EU has in recent years protected the “right to be forgotten”** for individuals who want to remove embarrassing search results about themselves from the public eye[25]. Users of Ubiquitous Voice Assistants may, too, soon realise they need such protection. It is undoubtedly convenient to be able to go back and find the exact date and time you met with someone, but you don’t want this information accessible to anyone else but yourself. Equally, when you delete a photo, it disappears from your view, but the photo remains on some servers, most probably in a different continent, “just in case” you want to retrieve it at some point, especially if you had accidentally deleted it in the first place. This creates a conflict of interest between our interest and intentions and those of the service

---

23 https://thenextweb.com/artificial-intelligence/2018/05/11/voice-assistants-could-be-fooled-by-commands-you-cant-even-hear/

24 https://advocacy.mozilla.org/en-US/privacynotincluded/why-we-made

25 https://www.pirateparty.org.uk/party-magazine/european-court-justice-google-ruling-gives-dog-bone

provider. All data from the point of opening an account or joining a service can and will be used "against" us until the end of time, in theory. The service provider carrying our data on its servers may have noble intentions and impeccable ethics but third parties accidentally or maliciously accessing it do not necessarily. This is a particularly sensitive topic this year with the **new GDPR regulations** just enforced by the EU to give EU citizens more oversight and control over their own personal data, whether private or not[26].

Conclusion

So, despite the desirability, richness and effectiveness of using our personal data as the background context against which Ubiquitous intelligent Voice Assistants can recognise what we say more accurately and react in a more individualised and relevant manner, a fine balance needs to be struck between personalisation and customisation, convenience, efficiency and seamlessness of experience, and safeguarding our data privacy and our personal digital and physical security.

[26]https://www.theguardian.com/technology/2018/may/21/what-is-gdpr-and-how-will-it-affect-you

Ubiquitous Health

By Charles Jankowski

Good Morning!

It's the week you've dreaded for months. Your doctor's exam is scheduled in a few days. You're really not looking forward to this, but you know that it's better for you. Okay, you know you have to do it. Now, let's see, you know there are some things you have to do to get ready, but what were they? The questions are running through your mind:

- Isn't there stuff I couldn't eat or drink a couple of days before?
- What about my meds? Oh right, when I left the doctor's office a few weeks ago, they handed me this huge ream of legalese that they said I had to read beforehand.
- Now where was that paper with the instructions?

This is how it works now. When you're about to have a procedure, you get this handout that probably has buried on the middle of page 5 small details like "no food or drink of any kind the day of the exam." You came home from your pre-test office visit and promptly stuck this multi-page document in one of various black holes of paper in your house. So when you need it, you don't know where it is. What happens if you either forget, or just don't worry about the instructions? No big deal, right? Wrong. If you show up for the test "unprepared," not having followed the instructions, they have to reschedule you; they can't do it today. So not only is your time completely wasted, the office loses that appointment slot, which of course only raises their (and our) costs more. More on those costs later. With all the technology at our disposal, there must be a better way.

There certainly is!

Speech and natural language technologies and systems, the promise of "ubiquitous voice," that we all work on and so passionate about, can not only help our users have better experiences, which in itself drives many of us, but can help move forward one of society's truly great challenges, that of improving our health.

Ubiquitous Voice

Can I drink coffee?
What can I eat today?
Can I eat pancakes four days before my procedure?

When you see *text in italics*, those are actual questions posed to an open-ended, natural language, intelligent assistant, which has been deployed with hospitals and clinics. The examples you see here are for the particular domain of the colonoscopy exam, which is a somewhat invasive procedure, this requires a fairly detailed preparation protocol involving what to eat or drink, limits on medications, a laxative you must drink beforehand, sometimes in two completely separate doses, and restrictions on activities. To make things even more confusing, the rules on food, drink, and medications change as you get closer to the exam. Take our first example of coffee. Two days before, you can have it without any limits. The day before, it's fine, BUT you cannot have any dairy with it (but sugar or any other sweetener is fine). Starting midnight before your exam, you cannot eat or drink anything, including water. And that's just coffee. Even if you found that handout, it's pretty confusing. And don't forget, if you don't follow the protocol, there's a chance that you'll waste your time and all that preparation, and you'll have to do it again.

So instead of the handout, wouldn't it be so much better if we can ask what we can eat, or whether we can eat something, like these patients did? You're in the kitchen thinking about breakfast, but you don't know what you can eat or drink? Ask! You're in a restaurant, and don't know what you can order? Ask! And this example system we're showing here knows all the rules, so "Can I drink coffee?" two days before will give a different answer than the day before, and yet another the morning of. As it should! And you don't have to remember the rules, or where the handout is, or any of that. If you could (from your Amazon Echo, voice-enabled TV or mobile phone) just ask what you want to know, and get relevant and time-sensitive answers, that is a much better patient experience. This is not only great if you are preparing for an exam, but lowers the complexity and the barrier to getting this important screening. I'm sure many folks read through the handout and either cancel or just don't show up due to the overwhelming amount of information. This is wholly unnecessary.

Not another Siri

> *How do I find out what time my appointment is?*
> *Can I still get the colonoscopy if I have a cold?*
> *Can I take my chemotherapy pills?*
> *Can I brush my teeth on the day of the exam?*
> *What if I throw up the bowel prep?*

Of course, it's not just food and drink you can ask about, there are questions about activities and medications, side effects and/or conditions, and just general questions regarding your appointment. Which brings up an important point. The very popular "general" intelligent assistants like Siri, Alexa, Cortana, and Google Home have very developed and intelligent methods for automatically mining the Internet for answers to questions, which is great, and very amenable to questions along the lines of "What is the population of Sri Lanka?" As the intelligent assistant space grows, and voice does indeed become more ubiquitous, we will see a proliferation of very specific personalized assistants, designed to help you perform a very particular task or engage in a very targeted activity. Personalization if of course key; to properly answer "Can I drink coffee?" the system needs to know who is asking, and then more importantly when that person's exam is, so the right context and time-sensitive answer can be given. But also, especially when getting into very sensitive areas such as health care where it is very important to give the right answer, there will need to be a whole process around not only generating answers to questions, but also having the provider (e.g., the doctor or clinic) approve those answers, so their patients get the correct responses. From both medical and legal perspectives, this puts a somewhat different spin on the process used to come up with the answers; auto-web-mining will no longer be sufficient. And this is largely why the design of this particular platform deviates somewhat from those of the more generic assistants; it is designed to take existing medical providers' handouts and create questions and answers from them, and have the provider approve the answers before deployment, and then leverage the bulk of existing content for new providers and domains. This surely sounds daunting, and requires a level of domain-specific knowledge that just is not necessary or desired for many of the current systems. But of course this seemingly high price comes with it the markedly higher utility of these specific assistants, as they are designed to help users get a very complex task done better. These specific personalized assistants really have the potential of improving the lives of our patients, our customers, our friends and family.

A Big Problem

Let's switch gears a little now and talk about not only how Ubiquitous Voice in health care can help the individual, but also the whole system. In 2015, Americans spent about $3.2 trillion on health care. An even more shocking statistic is that health care costs as a percentage of GDP, or the total economy, is now almost 18%; in 1960 it was 5%. The average family of four spends over $20,000 per year on health care; seven times what it spends on gas. I think that most people would agree that health care remains one of our Big Problems.

So far, we've talked about "acute" care, or prep/discharge for one single procedure, such as an exam or surgery. Potentially even more impactful is directing specialized intelligent assistants at "chronic" care, or ongoing conditions such as heart disease, cancer, bronchitis/emphysema, stroke, Alzheimer's, and diabetes. The numbers behind chronic conditions are staggering: according to the CDC, half of all adults have at least one chronic condition, and one quarter have two or more. Seven of the top ten causes of death are chronic conditions; the top two (heart disease and cancer) account for almost half of all deaths. From a cost perspective, 75% of health care costs are for such chronic conditions, many of which are preventable.

We close with this detour away from AI to make an important point. "You promised me Mars colonies. Instead, I got Facebook," lamented Apollo 11 moonwalker Buzz Aldrin. Now, I'm not at all trivializing the incredible explosion of our capability for people to be socially connected across distance and time, but it is particularly exciting for me to consider how, with health care, Ubiquitous Voice can help. And I do mean help. An AI system is not going to solve our healthcare crisis; Buzz' Apollo project took 400,000 people and $120 billion in today's money. But I can think of few domains where our ever-growing machinery of AI, as well as great design, can be harnessed more towards solving one of our Big Problems.

Are Friends Electric?

By Wally Brill

I've been embroiled in the debate about how much personality is too much personality for systems and devices that speak to us for fifteen years. There have long been two camps. On one side there are those who feel the machine should be... well, machine-like. Think Robbie the Robot from *Lost In Space* ("Danger Will Robinson! Danger!"). Then there are those who believe, as I do, that whatever the voice, people will create a clear picture of who they're speaking with within a couple of seconds and that they'll be more successful in their interactions if the system is highly conversational and very natural.

Given that we ascribe a variety of attributes to the voice we hear, we need to accept that the voices we interact with will simply become more and more natural and when driven by AI, surprisingly engaging.

That raises a number of questions.

- How will we select the voice to be appropriate for a wide ranging demographic and a broad remit?
- How anthropomorphic is just too much?
- Is there really an "Uncanny Valley"?
- As AI matures to the level of the Turing test and beyond, and my Intelligent Assistant knows all there is to know about me, what social implications will there be? Will I feel my sense of personal space, life and identity change?

Let's explore...

WE LIVE IN THE FUTURE

I talk to my robots a lot. Alexa handles my shopping list and plays music while I'm cooking. Siri sends text messages for me and my Mazda helps me call home when I'm stuck in traffic. But we're really in the stone age of where our relationships with speech enabled, intelligent assistants will go. As artificial intelligence improves and our interactions with devices become more and more sophisticated will we be satisfied with the sound of the voices speaking back to us?

The strong likelihood is that speech enabled entities will provide us with all the functionality of trusted helpers. They'll plan our trips, reserve us tables in restaurants, make purchases on our behalf and even do our taxes. They'll be the hubs of our home automation and that's only the start.

So if I'm interacting with these entities so frequently, what should they sound like when they respond to me?

PERSONA

The persona of any system is the consistent character. It's not just a voice: It's a combination of voice, interaction design and dialog. It's the impression the system gives of "who" you're interacting with.

In *Wired for Speech: How Voice Activates and Advances the Human-Computer Relationship,* Professor Clifford Nass from Stanford University says:

"...when people hear any voice, no matter how clearly not human, they automatically and unconsciously use their voice-analysis skills to assign a personality to the voice. Once they make this judgment, they may go even further and use the seeming similarity or difference between their perception of the voice's personality and their own to draw conclusions about the system's intelligence, likeability and trustworthiness, exactly as if they were interacting with another person."

That's precisely why people refer to Siri as "she." It's also why so much care must be taken to get it right. If I don't like the voice or if it annoys me, I'll use the system less frequently and it will be less engaged in my life.

DOWN IN THE VALLEY

On the other hand, have you ever seen a robot that's so lifelike it mimics human facial expressions? It can be initially impressive but when it makes a mistake and smiles in the wrong way or at the wrong time, the cleverness becomes uncomfortable and creepy. In 1970, Japanese roboticist Masahiro Mori first warned us of the "uncanny valley". That's where an automaton is so humanlike that we do a subconscious flip and begin to trust it to be consistently so. But when it appears that close to the real thing, any unexpected behavior, facial expression or speech disfluency can make us uneasy. Then we fall into the "uncanny valley". It's the space between obviously mechanical robots and near perfect simulacra such as the "replicants" of *Blade Runner*. The problem lies with the ones that are really close to human but just not quite right. The "uncanny valley" becomes less of a problem when there is no visual component and we're only dealing with speech. But we still need to avoid trying to fool anyone.

Speech recognition is really quite good now. Alexa even understands when I ask her "Where is Mt. Tamalpais?"; a place that is spelled quite differently than it sounds. (The intelligent assistant "Hound" thought I wanted "Car dealers in Mt. Temple Pints.") But though Alexa understands me perfectly, when she gives me the answer, she struggles with the pronunciation of the actual name. That makes me have to stop and think *"Did she get it right?"* It breaks the flow. That's an issue of text-to-speech. It relies on literal pronunciation. The other problem is prosody, the melody of speech. It's the pitch, stress and timing of the phrase, dictated by context and the speaker themselves. Text-to-speech engines still have some issues knowing how a particular phrase might be nuanced when spoken by a human. That's why I can't bear to have my emails read out to me or listen to robots reading the news. The mistakes of prosody make it very uncomfortable and the overall affect comes across as vacant and flat.

That said, huge advances are being made in the improvement of the technology and doubtless it will be relatively seamless soon.

IS THIS THE PARTY TO WHOM I AM SPEAKING?

So put on the shoes of the designer tasked with making recommendations for the persona for an intelligent assistant (SIRI, Alexa, Hound, Viv, etc.) What should it sound like? What should the persona be? Male? Female? Younger? Older? Then there's the question of register. Register is the level of formality someone uses when communicating. So, will the persona be familiar, like a peer or more distant like a servant or perhaps a being of superior knowledge and gravitas. Will it respond purely in text-to-speech or will it mostly employ recorded, concatenated speech with text to speech reserved for dynamic content? Remember you might be conversing with this persona many times a day. What about pace and verbosity. Will it be more succinct in its responses as it gets to know me?

I'LL FOLLOW YOU ANYWHERE

So how many of these robots will we have? Do we need more than one? Or will there be one universal interface that continues the conversation in my car, in my office, at home or anywhere I happen to be. Will it go so far as to speak on my behalf? (Voicemail systems already represent us.)

My robot continually learns about me, my life and my preferences. What kind of seat do I like when I fly? When will I run out of tofu and kale? If I'm in the kitchen packing the picnic basket and I say I want to go to Grandma's house, will it route me away from the Big Bad Wolf and will the navigation information be there in the car when I start my journey? (Don't forget to put the Huntsman on speed dial just in case.) Will my lights turn on as I enter the driveway when I come home and will the house already be warm? Yes, yes, yes and yes. Things become convenient and easy as we automate the regular and repetitive tasks. But I should only need one electric friend.

Things will get more complicated as enterprises employ speech enabled, intelligent agents to interface with customers while preserving their brands. Imagine Frieda the Freedonia Airlines surrogate, offering to provide ticket booking etc. Why would I want to work with her when my own robot knows so much about what I want and expect? The question of whether brands will allow my universal interface to connect to their systems is going to be an interesting debate.

WILL IT KILL ME WITH KINDNESS?

Then there's the question of how much help is too much help. At what point do I surrender the desire to do things for myself? What are the boundaries I set with my robots? Will my behaviors become templates that are enforced and repeated ad infinitum? Imagine advanced AI that can anticipate my needs coupled with the ubiquity of the interfaces that I live with. Will I begin to be overwhelmed and lose contact with tasks and choices that were previously part of daily life? Sometimes it's the randomness of things that can initially appear less convenient and less usual that provides extra color to mundane tasks. Perhaps I'll try crunchy peanut butter for a change? I can imagine getting to a point of telling my robot companion to just leave me alone for a while and stop taking care of me!

So the robot is here to stay and voice is the most natural and simplest way to interact with it. It'll be nearly perfect in its understanding and continually smarter. It'll be pleasant to interact with. It won't be trying to be a human but it will use all the conventions of conversation, speech and sound that will make it invisible as an interface. Alexa and her brothers and sisters are an exciting peek into where we'll go as the technology matures and they're networked and connected to "smart" everything. It's a good thing I like her voice.

Voice in the Vehicle: The Next Frontier

By Lisa Falkson

The time is fast approaching for voice to become the primary interface for computer-human interaction. Speaking to a mobile phone has become commonplace since the invention of Siri, and now the proliferation of in-home devices (such as Amazon's Echo and Google Home) is driving the adoption of speech recognition in the home automation market as well. While voice user interfaces have existed in the automobile environment for more than a decade, they have yet to break through as the primary mode of interface in the vehicle. The reasons for this are manifold: First, the automobile is a challenging environment for speech technology, with background noise impacting recognition accuracy. Second, the long production and lead times of the automobile industry mandate that the software and hardware are frozen years ahead of time; this automatically renders the speech systems out-of-date by the time they are released. And lastly, these systems have largely been designed in house by auto manufacturers themselves, without the help of voice user interface (VUI) professionals to design the interaction. This results in a lack of intuitive and conversational interface design, which frustrates users and curtails the adoption of speech systems.

What will it take for speech to become the primary interface in the car? To explore this, we need to address the issues that have prevented it from becoming widely accepted thus far.

> **Accuracy.** Automated speech recognition (ASR) is continuously improving, however there will always be challenges in accuracy. Proper names, for example, can be difficult to recognize without any additional information – therefore, for something like voice dialing, it works best if a

user's phonebook is uploaded and can be used by the recognizer to determine which name was spoken. Addresses (in particular, street names) will always be challenging as well, so using a database that contains supporting information is a requirement for improving accuracy. So, a large component of accurate speech recognition is in the software development process of the application; one cannot simply depend on the out-of-the-box capabilities of the recognition engine on its own.

Software/hardware constraints. While periodic hardware changes/upgrades in the vehicle are not realistic, software updates can and should be done. Therefore, the design of the software architecture must allow this flexibility. Ideally, the bulk of the recognition and natural language understanding is done on the server-side; therefore, if a change needs to be made – for example, adding a phrase to be understood – it can be done without any impact to the user. If a software update needs to be implemented locally (on the software that exists in the vehicle), those can also be remotely pushed directly to the car similar to upgrading an app on a mobile phone.

Design and Usability. In-car speech recognition systems must be designed ***by voice user interface professionals***. Even a simple command-based system that only handles phrases such as "call Joe" or "navigate to Starbucks" should still have smart error and help behavior in the cases where recognition can (and will) fail. The simpler and better the design, the lower cognitive load will be on the user.

The Problem

Back in spring of 2015, I shared this model for different types of driver distraction in the vehicle (see Figure 1).

3 types of driver distraction in the car:

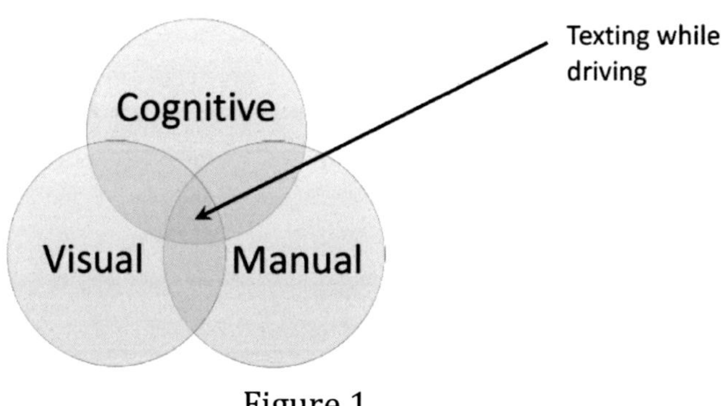

Figure 1

Visual = eyes off the road
Manual = hands off the wheel
Cognitive = mind distracted by tasks other than driving

Texting while driving, now statistically a commonplace act in the US and worldwide, is the intersection of all 3 types of distraction.

The design challenge for a vehicle's voice system is therefore to accomplish the following:

- Simplify the necessary tasks required in the car
- Respect the driving environment
- Keep eyes on the road, hands on the wheel
- Minimize driver interaction with screen
- Use speech recognition input and TTS output for majority of interaction
- Optimize pre and post-driving information on the screen

In 2016, I presented the following information at Mobile Voice Conference in San Jose, courtesy of the Wall Street Journal (see Figure 2):

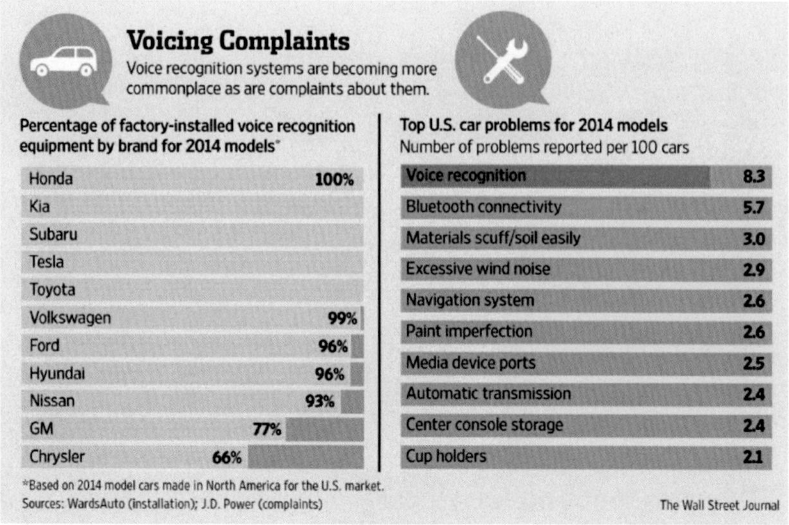

Voicing Complaints

Voice recognition systems are becoming more commonplace as are complaints about them.

Percentage of factory-installed voice recognition equipment by brand for 2014 models*	
Honda	100%
Kia	
Subaru	
Tesla	
Toyota	
Volkswagen	99%
Ford	96%
Hyundai	96%
Nissan	93%
GM	77%
Chrysler	66%

Top U.S. car problems for 2014 models Number of problems reported per 100 cars	
Voice recognition	8.3
Bluetooth connectivity	5.7
Materials scuff/soil easily	3.0
Excessive wind noise	2.9
Navigation system	2.6
Paint imperfection	2.6
Media device ports	2.5
Automatic transmission	2.4
Center console storage	2.4
Cup holders	2.1

*Based on 2014 model cars made in North America for the U.S. market.
Sources: WardsAuto (installation); J.D. Power (complaints)

The Wall Street Journal

Figure 2

Unfortunately, little has changed since this data came out in 2015; this is typical of the automobile industry, which moves fairly slowly compared to the technology industry (particularly mobile devices and software).

Partnering for Success

One defining change in the industry over the past few years is the reduced resistance to partnerships. Apple Carplay and Android Auto opened the door to major OEM's for integration with highly accurate, cloud-based recognition platforms. However, the integration itself can become an issue, if not properly executed.

In Figure 3, data shown for independent study by AAA Foundation for Traffic Safety, shows that in-car systems still have moderate to high distraction ratings.

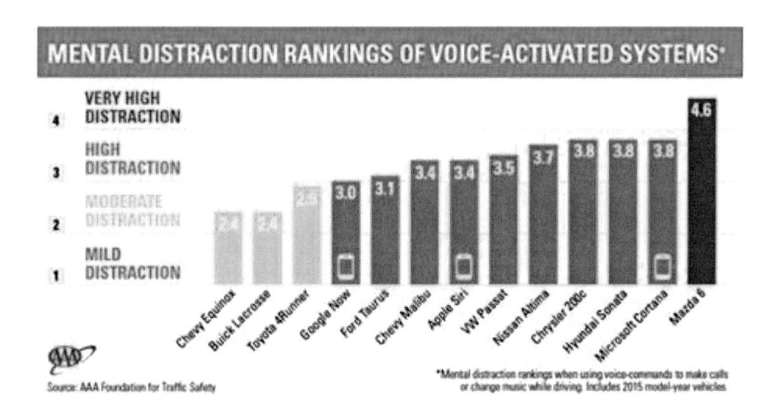

Figure 3

Despite attempts to make voice-activated systems that are hands-free and therefore safer on the road, the levels of mental distraction were still very high. This includes both "Google Now" and "Apple Siri", along with HMI systems from top auto manufacturers.

Further data from JD Powers shows:

- 19% of OEM GPS navigation users were unable to locate a desired menu or screen
- 23% had difficulty with voice recognition
- 24% claimed that their devices provided incorrect routes.

Specifically,*"high level of integration makes them incredibly convenient, but it has also led to usability issues. According to a study performed by J.D. Power and Associates, most consumer complaints about OEM navigation systems are related to ease of use."*[27]

---

[27] Source: Lifewire, October 16th (https://www.lifewire.com/oem-infotainment-systems-navigation-534746)

Regardless of how well the speech interface itself works (and it often doesn't), there is little use if the driver can't find the button to initiate it, or has to navigate deep into menus to find it.

When drivers feel that their car's system is out of date or difficult to use, they turn to the obvious: their mobile phone. Chances are high that they will have a modern device in their possession that is less than 1.5 years old. That device will either have Siri or Google Assistant - or a custom mobile app that uses speech recognition and/or TTS in (such as Waze, HOUND, Melody, or others).

All the top "assistants" with the highest accuracy are an easy alternative to the built-in system built by the auto manufacturer. Therefore, the built-in system is often obsolete before it has even launched with a new car.

Rapid Growth = Demand for Change

As the connected car industry grows over the next several years, there are great opportunities to improve safety, convenience, customer experience, and quality of life. The projected numbers are astonishing:

- Connected vehicles will outpace population growth for the next decade.[28]
- IHS Automotive forecast: 152 million actively connected cars on global roads by 2020.

If this is truly the case, then the connected car industry is poised on the brink of explosive growth. This suggests that the demand for more highly useable voice-driven interactions will increase as well, since they are the weak link in the current user experience.

The trend towards better in-car voice systems should be similar to the increased interest in "highly usable" websites during the dot com

[28] Source: http://bigdatanomics.org, November 2016 and http://sas.com whitepaper: "The Connected Vehicle: Big Data, Big Opportunities"

boom of the late 1990's/early 2000's, or the (belated) emphasis on "good user interface" for mobile applications on smartphones from 2007 onwards. While there are still poorly designed websites and inefficient mobile apps, companies are increasingly subjected to scrutiny by users when they provide a poor user experience.

Voice and Autonomous Vehicles

With the predictions for autonomous vehicles becoming available within the next few years, some might ask why voice systems will still be important. After all, if the car is driving itself, hands-free is no longer an issue. However, there are still many says in which a voice system will play an important role:

- Communication between user and vehicle regarding the autonomous/manual driving modes. The user should be able to change modalities via voice, and also have the vehicle communicate very clearly about the mode of operation it is currently in.
- In non-autonomous driving modes, the hands-free functionality will still be important.
- An autonomous driving situation becomes closer to the home environment, where hands-free is still a luxury (for example, setting a timer with Alexa in your kitchen). If hands are busy with another task, the user will still enjoy being able to use a well-designed, accurate voice-based system.

The Future

Voice is clearly the interface of the future, and vehicles will be no exception. My prediction is that increased partnerships as well as demand from the industry will drive the development of more highly accurate, usable systems. However, as with everything in the car industry, these changes will come more slowly than they have in the standard consumer electronics industry.

Who Are You?

By Phil Shinn

The Problem

How do we get recognized and authenticate in a world without keyboards? How will some car, or Echo, or bathroom mirror[29] know who I am, and tell me my stuff, and not give it to someone else?

My car needed work and the garage gave me a loaner. The first thing the loaner car wanted to know after I turned it on was my contact list.

My Echo responds to me or my 3 year old when we ask "Alexa, what's the weather today?" Will Amazon accept and fulfill the request "Please send me 25 Nintendos and 100 My Little Ponies" from anyone in earshot?

In the future, when we share stuff like cars (I love Zipcar) or dwellings (I love AirBnB) or networks (I love Starbucks) a lot more, the challenge will grow.

Other Solutions

I'll start off by being up front – I'm a speech geek. So take what I say with a grain of salt. But I hope by the time you finish this section you'll reach for the vodka and lime.

Mo Bettah Passwords

Passwords are ancient and vulnerable. Making them more complex so they are less guessable only makes them less usable. Cracking vulnerability is the least of it. Ever type a password when using a public internet? Sure you don't. Even at home, key-loggers installed with the click-on-an-attachment render those complicated passwords so much toast. So, you have an iPhone, so no problem, right? [30]

---

[29] http://www.glancemirror.com/

[30] https://www.washingtonpost.com/news/the-switch/wp/2016/08/25/this-malware-sold-to-governments-helped-them-spy-on-iphones/

SMS Call Back 2FA

Some time ago using two factor authentication (2FA) on a mobile device using SMS became standard. That it were secure.[31] The attack surfaces are multifaceted, the simplest being social engineering SIM swaps.[32] The problem is using SMS call-back 2FA isn't really leveraging 'something you have' but instead is 'something they sent you.' In addition to social engineering, it's vulnerable to man-in-the-middle attacks that can be initiated with very inexpensive software[33]. No need to get hold of a Stingray.[34] So while 2FA SMS beats passwords alone, it is NOT secure. NIST writes "...using SMS is deprecated, and will no longer be allowed in future releases of this guidance."[35] See also [36] and [37].

KBA: AKA Authentication by Interrogation

So you try to log on to some website you haven't visited for 5 months and you don't remember the password. You call them up, get cheesed at the IVR and pound out to some poor schlub working for minimum wage.

They start asking you "Can you tell me your phone number?" WTF? Which of the dozen numbers I can't recall did I use back when? Are you making this up?

---

[31] "So Hey You Should Stop Using Texts for Two-Factor Authentication," Andy Greenberg, Wired, June 6, 2016.

[32] https://www.dos.ny.gov/consumerprotection/scams/att-sim.html

[33] "Top 5 Apps to Spy on Text Messages – SMS Tracker Reviews," http://safeguarde.com/

[34] https://en.wikipedia.org/wiki/Stingray_phone_tracker

[35] "DRAFT NIST Special Publication 800-63B Digital Authentication Guideline," NIST, https://pages.nist.gov/800-63-3/sp800-63b.html

[36] "NIST is No Longer Recommending Two-Factor Authentication Using SMS", Schneier on Security, https://www.schneier.com/blog/archives/2016/08/nist_is_no_long.html

[37] "Time is Running Out For This Popular Online Security Technique," Fortune, July 26, 2016, http://fortune.com/2016/07/26/nist-sms-two-factor/

"What was your high school mascot?" Seriously? Anyone with an internet connection can get that in about 10 seconds, or they have it already if they're using FacebookStalker [38].

What was my street address two addresses ago? Last four digits of my social security number? My mother's maiden name. Seriously? You think I would actually say that out loud?

Back in the day when this all got started, it was tough to get this information. You needed to go to the courthouse and pull papers. Not anymore.

And even if that information is not public, but instead is in 'the enterprise' (ooooh) what makes you think that's safe? How many credit card and mortgage companies and whatever national databases have been plundered? If you worked for the US government, your name, address, employment history, medical records, employment evaluations and even your fingerprints got hacked [39] and are probably for sale. If you paid taxes online, you are vulnerable.[40]

Just as NIST weighed in on SMS 2FA, the FFIEC, the government agency that regulates financial services, has given guidance that KBA should not be relied upon: "Institutions should no longer consider such basic challenge questions ... to be an effective risk mitigation technique." [41]

So, KBA is a pain AND insecure. To add insult to injury, it is expensive. First you pay that contact center agent for the time to ask the silly

---

[38] "FBStalker Automates Facebook Graph Search Data Mining", Threatpost, Kaspersky Labs.
[39] "Hacks of OPM databases compromised 22.1 million people, federal authorities say," Ellen Nakashima, The Washington Post, July 9, 2015.
[40] "IRS Breach Highlights Weakness of 'Knowledge Based Security'", Zach Noble, FCW, May 27, 2015,
[41] Federal Financial Institutions Examination Council, June 22, 2011.

questions they don't actually know the answer to but which they are reading, and take the caller's heat, but you also pay the information provider a fee per transaction to get the information.

Biometric Canards

They Cannot Be Revoked

So the bad guy steals your fingerprints, or gets a recording of your voice, or a picture of your face, whatever, if you are not wearing a hijab. Security is compromised, and cannot be recovered, because the user can't change to a new set of fingerprints or face or whatever. Unlike PINs, passwords or crypto-keys, biometrics cannot be revoked, or so the story goes[42].

I was surprised to hear this from the lead security architect at a large financial institution where I once worked. "Oh, biometrics, no good, they can't be revoked." He smirked. I wondered how he got his job, and how long he would keep it.

The obvious counter measure is good liveness-detection, necessary other reasons (see below). In the contact center, where text independent speaker recognition is being deployed, the agent is speaking with the caller, and some contact center agents have pretty good liveness detection.

That aside, more complex methods produce cancelable biometric templates, and these have been around for years. See, for example, Patel et al.[43], who catalog more than 20 methods. See also [44]. Any

---

[42] "Why fingerprints, other biometrics don't work," Dave Aitel, USA Today, Sept. 12, 2013.

[43] "Cancelable Biometrics: A Review," Patel, V. et al., IEEE Signal Processing, Vol. 32, Sept. 2015, pp. 54-55.

[44] "Cancelable Biometrics," Jin, T.B. & Hui, L.M., Scholarpedia, 2010.

security professional who doesn't know what a biometric salt is isn't credible.

They Are Not Persistent

Funny hearing this one from the password posse. How often do I need to change it now, every 2 weeks? And I have to use upper and lower case, some numbers, some alphas, some other stuff and an emoji? Or you lock me out of the network until I speak to someone who asks me the last four digits of my SSN?

They Are Not Foolproof

Nothing is. Deal with it.

They Mean You Can't Take the Fifth

Yes, if you are worried that you could be compelled by law enforcement to give up your fingerprint, unlocking your kiddie porn, but you cannot be compelled to give up your password (that's speech and protected), then yes, rely on a password.[45]

They Can Be Spoofed

Poor liveness detection results in spoof-ability. iPhone TouchID was hacked shortly after its release.[46] Some biometrics are laughably weak. [47] Given the critical nature of liveness detection, researchers are working hard on counter-measures. [48]

One dynamic passphrase system enrolls you saying the digits zero through nine three times[49], and then authenticates[50] by prompting

---

[45] "Apple's Fingerprint ID May Mean You Can't 'Take the Fifth'", M. Hofmann, Wired, Sept. 12, 2013.

[46] https://www.ccc.de/en/updates/2013/ccc-breaks-apple-touchid

[47] https://www.youtube.com/watch?v=joQJqq7DW1U

[48] http://pralab.diee.unica.it/sites/default/files/Akhtar_PhD2012.pdf

[49] https://www.dropbox.com/s/ma1flyipl5mtuzn/VKOP%20Enrollment%20Demo.mp4?dl=0

[50] https://www.dropbox.com/s/n1gr4qo5g1qa2z1/VKOP%20Authentication%20De

with a random five digit number, foiling a recording replay attack. The multi-modal version also does simultaneous face-match and has a model of lip movements that correspond to digit strings. So if someone put up a selfie and re-played the audio of the digits, liveness detection will fail.

Conclusion

"On the Internet, nobody knows you're a dog."

Figure 4- Peter Steiner, The New Yorker, July 1993

In a world without keyboards, passwords have to change. Using the audio of text-dependent passphrases that require enrollment and authentication with the same phrases are better than nothing, but are vulnerable to a replay attack. You can raise the security level by using dynamic passphrases, challenging the user to say something different with each authentication. But once the bad guy is in he's in.

---

mo.mp4?dl=0

Continuous authentication is possible using text independent speaker recognition. The amount of audio required is more, but in the context of speaking with your IoT devices, this is natural. The next challenge is finding counter-measures to detect the spoofing of speaker verification systems using text to speech synthesizers/modulators and similar technology.[51] [52]

---

[51] "ASVspoof 2015: The First Automatic Speaker Verification Spoofing and Countermeasures Challenge," Wu., Z. et al., Interspeech, 2015.
[52] "STC Anti-spoofing Systems for the ASVspoof 2015 Challenge," Novoselov, S. et al., ICASSP, 2016, SP-P5.

But That's Not How People Talk: Making Your VUI More Human

By Cathy Pearl

What do the Amazon Echo, Google Home, Siri, Cortana, Hound, the Comcast Xfinity Remote Control, and my Toyota Matrix all have in common? To talk to them, *you* must initiate the conversation.

It's an interesting change from the early days of voice-enabled technology. In the late 1990s, IVRs (Interactive Voice Response) became a common way to communicate with companies. The "conversation" began as soon as the call was connected, and the caller spoke to the automated system. Back then, the user was led through a series of voice interactions, collecting the necessary information to complete a task.

Nowadays, most voice assistants are one-offs: the user requests an action, and the voice assistant attempts to complete it. Although devices like the Amazon Echo (and now Google Home) are using this "one-turn" method with great success, in a way, we've thrown the baby out with the landline. IVRs got a bad rap, with websites like "Get Human" popping up, and skits on sketch comedy show Saturday Night Live poking fun, but IVRs did certain things very well, and very successfully. It's worth examining some of the conversational principles IVR designers created, and bringing the best of these into the new era of speech recognition technology.

What IS Conversational?

First off, it's important to define what "conversational" really means. It's a word being used quite frequently in relation to "bots" and VUIs: the promise is that soon we'll be able to have a conversation with our fridge, our car, our watch, and our thermostat. But there are very few truly conversational systems out there.

To be conversational, a VUI must have:

- **More than one "turn" (e.g. more than one interaction, in a row, with the system)**. If I came up to you on the street and said "Hello, how are you?" and you said "Fine" and walked away, we would not call that a conversation.
- **A memory of what's happened in the conversation up to that point**. Imagine a shopping bot. I've requested a black sweater; later I ask to see the same sweater, but in blue. Many bots today can't handle this request.
- **The ability to recognize pronouns**. This is something even toddlers can do, and it drives users crazy when VUIs fail to understand a simple pronoun such as "it".
- **Allowing for multiple ways to refer to the same thing** Not only does a conversational VUI need to recognize different ways someone might refer to the same item (e.g. "sandwich" vs "hoagie"), it also needs to understand that people give information in different chunks. Sometimes the info is all up front ("I want a large pepperoni pizza") and sometimes it's in multiple turns ("Yeah I'd like a pizza please.")

Now that we've defined some standards for whether or not something is conversational, what's the best way to make it clear *within* the conversation when it's the user's turn to speak?

Command-and-Control vs. Conversational

There are two ways to communicate with a voice user interface today. One is a "command-and-control" method, and one method uses natural human turn-taking cues.

"Command-and-control" refers to anything that requires the user to initiate the conversation at every turn. Things like Amazon Echo, Google Home, and the home robot Jibo require you to say a "wake word". Phone assistants such as Siri, Google, Cortana, and Hound also require either a wake word, or, for the user to press a button or a tap a microphone icon. In most of these cases, you must say the wake word or tap the mic every single time you wish to speak (although some of these are starting to tackle the challenge of longer dialogs, in which you can continue the conversation without having to say the wake word again). Amazon Echo and Google Home now have user-enabled modes that will leave the microphone on for a few seconds at the end of each turn, and Hound allows the user to simply say "Ok" as a shortened wake word if spoken immediately after the last turn.

Rather than think about the simple, one-turn examples that are so prevalent in today's voice assistants, let's take a page from IVRs, and instead start building multi-turn experiences.

One example of an extended VUI conversation that does *not* require wake words is from the company Volio (where I was a Principal Interaction Designer). Our iPad app created engaging, voice-driven conversations with pre-recorded video of an actor. The actor's face/upper body took up most of the screen, with a small 'picture-in-picture' box in the upper right-hand corner, showing the user's face (similar to Facetime). Figure 1 shows a screenshot. The example shown is from an app created with Esquire Magazine, in which the style columnist, Rodney Cutler, gives advice about hair products.

Figure 1: Volio's "Talk to Esquire" app

After Rodney asked a question (such as "How long is your hair?"), it was the user's turn to speak. This was indicated by the box around the picture-in-picture turning green. When the user finished speaking, the green box would fade, and the actor would continue the conversation, with different videos for different responses. The end result was a seamless back-and-forth conversation between the user and the app. (To make it seamless took a lot of care; the actor's head position and lighting was kept at a constant, to avoid any jumpiness between turns.)

We started building this app in 2011, and despite voice apps not being as prevalent, people had no trouble figuring out how to use it--even when they weren't told to speak. There were two main reasons for this:

- The interface was simple: just the face of the actor speaking and the picture-in-picture
- It was evident when it was time for the user to speak, because (a) the actor asked a question, and (b) the frame around the user's face would light up

People respond naturally when someone (or something) asks a question. We've seen older users who are less familiar with apps automatically respond to a spoken question by speaking.

Error Recovery with Volio

As anyone who has worked with speech recognition knows, even with today's advancements in the technology, it will still fail at times. After all, human-to-human speech often fails as well.

In the IVR days, when the user spoke and was not understood, a typical response would be "I'm sorry, I didn't understand. Please tell me the city you're traveling to again."

This can be rather cumbersome and annoying. At Volio, we realized we could take advantage of the fact we had a human face involved, and fall back on human conventions. For the first couple of errors, the actor doesn't say anything about the problem: he or she simply continues to look at the camera in an "active listening" mode. We quickly discovered people automatically repeated themselves, often clarifying what they had said. Using this method alone got us to recover from over 80% of our errors. Often, users did not even realize they'd been in an error state.

If multiple failures took place, the app would eventually back off to a GUI method, displaying some of the possible answers available, and disappearing again after the user tapped one. We never wanted to abandon a user if for some reason speech recognition wasn't working well.

Multi-Modal Conversations

Sensely uses a similar approach at with their virtual nurse Molly. Molly speaks to the user, and the user can speak back. Sensely's app is more complex, because although at times the only way the user can respond is via voice, at other times the user is able to respond by voice *or* with a graphical input, such as a button, list, or other widget. Because the types of responses are mixed within a single conversation, it's even more crucial to be clear when it's the user's turn to respond.

Sensely uses the same methods that were used at Volio: make sure the avatar says something (a question or an instruction) that makes it clear to the user it's their turn. In addition, it only turns on the microphone icon (there is no picture-in-picture in this case) when it's the user's turn to speak. For error recovery, in a speech-only state, it does use prompts to guide the user; in a mixed-mode state, the backoff is already available on the screen.

Example of the first couple of turns in the Sensely Daily Check-In:

MOLLY
How are you doing today? [*Voice only responses*]

PATIENT
I'm feeling pretty good.

MOLLY
Thanks for sharing. Let's get started. Do you have your blood pressure cuff and scale ready to go? [*Voice OR buttons*]

PATIENT

[Presses 'yes']

It should be noted that neither the Volio nor Sensely apps allow for barge-in. ("Barge-in" refers to whether or not the user is allowed to interrupt.) Amazon Echo and Google Home don't allow for barge-in either, *except* for the wake-word.

There is some disagreement about whether the user should be allowed to "fast-forward" some of these more directed conversations, and be able to barge in while the app is still talking. My philosophy is that if your user is frustrated and wants to move more quickly, the design needs to be reworked. It may be that your user is an expert who is still being given longer prompts more suitable for a beginner user. It may also be that your VUI is unnecessarily wordy. There are times when it's better for the VUI to display information than to say everything out loud (providing you have a multi-modal system).

As an example, the Sensely virtual nurse sometimes asks a question that has a list of possible responses, such as "Do you currently have any of these medical conditions?" Rather than read out a list of 10 of them, which is quite slow (as well as difficult to remember), she only says the question part out loud, and the responses are all displayed on the screen. At this point, the user can say the one they'd like, or tap it.

Cues to let users know when it's their turn

Regardless of having an actor, an avatar, or just an app, it's crucial to let the user know when they should speak. Here are some effective methods:

- Ask a question ("Did you sleep well?") or provide an explicit instruction ("Please put on the blood pressure cuff, and say 'Continue' when you're ready.")

- Enable microphone / other indicator (in the case of non-visual VUIs, such as the Echo or Google Home, it uses lights to indicate when the device is actually listening)
- Make it clear what the user can say (Don't ask "How can I help you?" if the only thing your VUI can actually do is book a hotel room.)

Some of these design principles are beginning to be enforced at the development level. Google, for example, will reject a submitted "Action" if the developer turns the microphone back on without making it clear what the user should say.

Home Assistants: Conversational Examples

Amazon Echo and Google Home are both beginning to expand their conversational dialogs. Here are some recent examples.

ME
Ok Google, what's my husband's name?

GOOGLE HOME
You haven't told me that; what's his name?

ME
Chris.

GOOGLE HOME
Ok, I'll remember that.

Note that I did not have to say "Ok Google" before I gave the answer; it continued to engage in listening, and this was evident because of the lights still being on. This is a terrific example of the assistant anticipating the next step in the conversation.

Here's the same question, but answered by Amazon Echo:

ME
Alexa, what's my husband's name?

ALEXA
I'm a little concerned you don't know that.

I have to admit this response made me laugh, but Google's follow-up was much more useful. It *learned.*

Here's another good example, this time from Amazon Echo, where I ordered cat food for our aging kitty:

ME
Alexa, order some cat food.

ALEXA
Based on your order history, I found two matching items. The first is Wellness Core Natural Canned Grain Free. It's $43.59 total including tax. Would you like to buy it?

ME
No.

ALEXA
Based on your order history, I found 2 matching items. Amazon doesn't have the cat food you previously bought. I found a similar product, Royal Canin Vet Calorie Control High fiber. It's $60.99 total. Would you like to buy it?

ME
Yes please.

ALEXA
Ok. Order placed.

Again, I did not have to say "Alexa' while we were still *within* the conversation. This one worked pretty well; it even proactively offered an alternative product when my previous one was not available. One quibble is that it was a bit stilted; a human would not have said "based on your order history, I found two matching items." Instead, it could have short-cutted with something such as "The other item, Royal Canin, is no longer available. I found a similar product..." But all in all it was a satisfying, and more importantly, easy process.

Conclusion

There's a good reason things like the Amazon Echo and Google Home require a wake word: you're often talking when you're around them, but you're not always talking *to* them. Imagine a world where, every time you said something, your home assistant tried to jump in and help. That would get old quickly. Instead, they wait patiently to be summoned.

However, there is a lot of room for what happens next. When I ask my home assistant about a birthday party my son is going to, for example, I want the ability to ask for more detailed information. Right now, I have to initiate a whole new conversation; it's as if we've never spoken. Users have control of when a VUI conversation begins; now they need control over when a conversation ends.

ivee Story

By Jonathon Nostrant

The Beginning

Life works in mysterious ways. If you asked me when I was a kid if I would have found myself dedicating 10 years of my life to speech recognition, I would have probably said no. However, you sometimes can't help whom (or what) you fall in love with. In many ways I was lucky. In my twenties, I fell in love and found myself on an exciting entrepreneurial journey building a company that developed speech recognition products, similar to Amazon Echo and Google Home.

The journey began when I first arrived in Hong Kong in January 2008. I decided to study abroad for a semester and this was my first trip across the Pacific. Luckily, I had my dad with me to show me the way.

Growing up, I had always admired his entrepreneurial spirit. Our family business made and sold consumer products, everything from colored extension cords to children's bath toys and Dad traveled back and forth to China often, always returning home with an exciting new product. Turns out, he too needed to be overseas at the same time.

Like most sons, I wanted to be just like Dad, so a 6-month study abroad program felt very fitting at the time. As people often say, timing is everything, and it just so happened that we stumbled into speech recognition on our first trip together.

Seeing is Believing

Edgar, a friendly local, with large glasses and a smiley face, ran a small company in Hong Kong called Cyberworkshop that engineered toys and consumer products. He greeted us at his entrance while his team worked diligently in the shop. They were typing away on computers that were surrounded by parts from prior creations, including the well-known children's toy, Furby.

"What you're looking at here," Edgar said, "is a development kit from a company called Sensory from Silicon Valley. This technology is Natural Time Set and when you speak to it, it will allow you to set the time and the alarm from just the sound of your voice."

Without a moment's notice, Edgar pressed the button on the kit and a small robotic voice on the speaker said, "tell me the current time."

All of a sudden, the engineers stopped typing and the workshop quickly quieted down allowing Edgar to say "seven... thirty... three... am." Then voila "7:33 am" appeared on the display!

Wow! This was the first time that I had actually seen speech recognition work and work well. I then proceeded to try it myself with the correct time (4:32 pm), and it worked again!

The Lightbulb Moment

People often ask entrepreneurs: "What was your 'aha' moment? When did the light bulb turn on and when did the vision unfold before you?" For me, this didn't really happen instantaneously. There was no exact moment when it clicked; I had to get my hands dirty first and then I started to see a vision for what the future might hold.

During my semester abroad, Dad continued to work with Edgar to finish an early prototype for a voice-activated alarm clock, which we later named Moshi.

Moshi Classic Alarm Clock

"Hello Moshi"

Command Please

"Set the current time"

What time would you like to set?

"Seven Thirty Three AM"

The time is now set to 7:33am.

It took about a year to finish development. When Moshi was ready you could ask her 10 simple voice commands (like "Time," "Set Time", or "Set Alarm") that were all initiated by a hands-free voice trigger, "Hello Moshi." Moshi truly was the first consumer hands-free speech device that was about to hit the market.

I was determined to get involved and help, but young and without many skills, I jumped in headfirst and Photoshopped the packaging for Moshi. Later, I blossomed into a product role, but I had no idea where to start so I did anything I could to help.

Before Siri

The year was 2009 and we now had a finished product to sell. There was no Alexa, Bixby, Cortana, Google Assistant, or even Siri on the market! The world was ripe for disruption.

To get the word out we prepared a press kit that we sent to several news outlets. We got lucky and Good Morning America picked up the story. They featured Moshi on live television at CES 2009 and the press fell in love with the concept of a voice-activated talking alarm clock.

Moshi sets the time by voice on Good Morning America at CES 2009

Voice, as we now know today, is an easier and simpler interface than touch. Not only adults, but children, teens, the elderly, visually impaired and the disabled all loved to interact with Moshi and her personality. Moshi became an overnight hit and we sold close to a million pieces in less than 24 months. It was a very fun and exciting time!

Hello ivee

Then something happened. They say in every failure there's an opportunity, but at the time this felt catastrophic. My parent's divorce led to a complicated family dynamic, which made me decide to leave the family business.

Frustrating and challenging as it was at the time, looking back this turned out to be a smart decision that lead to the birth of ivee.

Different from Moshi, ivee was going to fulfill a vision that unfolded during the Moshi experience. What we learned is that people wanted to do more than just set the time and the alarm with their voice. Our customers told us that they wanted a hands-free command center experience on their nightstand. Instead of just being able to say 10 simple commands, they wanted to be able to ask almost anything.

The best example that I could think of to describe the vision was from the movie Iron Man. Remember J.A.R.V.I.S. the invisible intelligent assistant? That's what we wanted to create.

Getting Started

Trying to get close in on the J.A.R.V.I.S vision, we chose a newer chip and technology that would respond to 50 voice commands, instead of the original 10 that Moshi had.

We quickly developed a new ivee product, but the world moves fast too. As we were preparing to launch ivee, Apple announced Siri.

Overnight, people were naturally talking to their phones and didn't need to use specific voice commands. Our technology was already outdated and I learned right then and there that **timing is everything**. We still launched the product and sold units, but we didn't get the initial pop and the traction that Moshi did.

Leaping Ahead

Even though Siri was now on millions of devices, we still had our vision. It was inevitable that voice would be the future interface and we would be talking to devices (other than phones) in our homes.

The challenge was: how could we get Siri-like voice technology into our product? We knew that we would need to connect ivee to the Internet.

In 2013, we launched a new product on Kickstarter that helped us raise awareness and sell pre-orders. The ivee Sleek was the first truly hands-free internet connected speech device that would respond to open-ended commands, not just a fixed set.

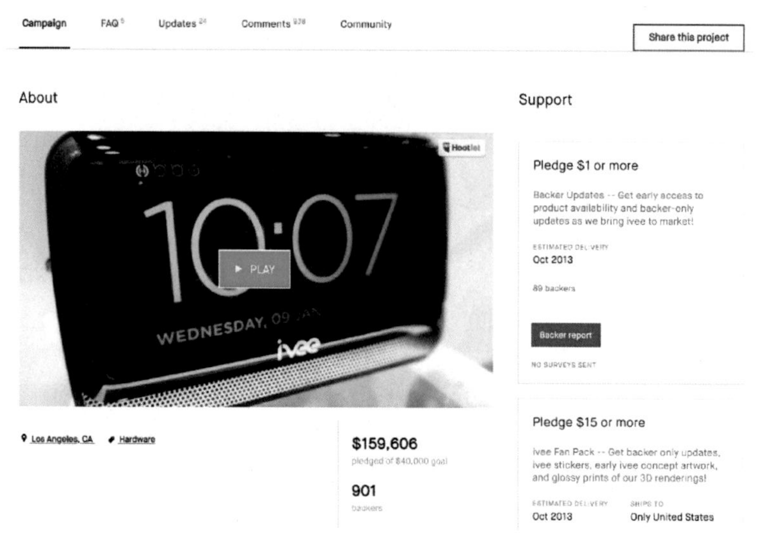

Screenshot of the ivee Sleek Kickstarter Campaign

Bootstrapping a product like Sleek to market was a tremendous effort and also a strategy that I would not recommend. We knew that we needed a financial partner so we met with Venture Capitalists (VCs) and Angel investors (angels), only to be turned down meeting after meeting. We learned that **most investors do not like to invest in hardware companies.**

"Why do I need an alarm clock or another device in my home listening to me, if I have Siri in my pocket?" This was the question that we repeatedly heard from investors. We countered with the explanation that your phone is mostly in your pocket, so why would you want to have to pull it out every time you wanted to talk to your assistant.

This didn't stop us. We pushed forward and launched the product without raising money and successfully sold the product to major retailers, including Best Buy and Lowes. The new company generated over $1M in sales in 2014 and we did this without raising any outside capital.

A Hard Lesson

Bringing ivee Sleek to market without the proper resources was the hardest lesson learned on the journey. We never should have released a product that wasn't ready for prime time. As a result, a majority of the product that we did ship ended up coming back to us. Because of this, we lost time and money.

Recognizing it was time to make a change, we applied to a start-up accelerator, which we believed would help us raise money so we wouldn't make the same mistake twice.

Even though we were unsuccessful in launching the new product, the traction that we did achieve allowed us to get accepted into a prestigious Silicon Valley accelerator, 500 Startups.

Accelerating Forward

Right before we were accepted, the Amazon Echo with Alexa was announced. The mentors at 500 Startups were supportive of our vision and they were not dismayed that Amazon was getting into the space. In fact, our traction in 2014 and Amazon's early success with the Amazon Echo allowed us to tell a unique story, which helped us raise over $1M dollar angel investment. Using the money we started to redesign ivee.

The ivee team at 500 Startups Demo Day, May 2015

Heating Up

Google then announced their version called Google Home and others started to announce their iteration of a voice assistant device for the home. Voice was becoming the next platform.

We started to see the writing on the wall. We knew that the space was heating up and was starting to become a "big boy" game. Everyone was talking about voice in 2016. We knew it was going to take more money to get a quality product into the market so we decided to do another round of investor meetings. Now they felt we were too late. First, they couldn't see it happening, and now we were too late! It was frustrating at the time, but pushed us down a new path: finding a strategic partner.

Corporate Dating

Over a period of 6 months, we met with 40 different companies, narrowed it down to a handful of interested partners, and ended up getting an official offer. It was a long, intense process.

After many conversations, lots of demos, and even some Jedi mind tricks, we selected a partner (whose name is omitted for confidentiality) who we felt would be the best company to help us take the new product to market. We signed a letter of intent (LOI) with this partner in the summer of 2016 and we were looking forward to joining their team in the fall.

After engaging in exclusivity, we began the diligence process which for us was 60 days (and could have been shorter). The process involved bi-daily calls with overseas teams, multiple code reviews, and lots of legal work. We were 10 days from closing and the partner unexpectedly decided to pull the plug for some non-obvious reason.

We later found out that the company that was in the process to acquire us was acquired themselves by an even larger company!

Entrepreneurial Learnings

Our deal fell apart and this felt like a failure at the time, but looking back on it now it actually wasn't. Although we didn't get the "exit" that we were hoping for, we were acqui-hired back into the industry to work on the same problems that we started with and we learned some very valuable lessons along the way.

1. **Market timing is everything**. Being too early or too late can determine your fate. It's important to get to market as quickly as possible with the best product as possible.
2. **Investors prefer software over hardware**. The hardware business is difficult and investors tend to shy away from these deals. These businesses typically require more capital to develop a quality product and are harder to iterate once a

product is in the market. It's generally harder to fund a hardware company, especially if you're a first time founder. Only build hardware if you must - and if you can't raise the money that you need, then don't build it.

3. **Don't launch a product prematurely**. Test your product with small test groups as much as possible before you launch and if your product is not ready, don't ship it to meet a rush deadline. You're better off not shipping at all. Customer dissatisfaction and returns will cost valuable time and money.
4. **Focus on timing over price**. If you get an LOI, focus on the time that it will take to close the deal. None of the other terms will matter if your deal doesn't close. We learned the hard way that time kills deals.

"Success is going from failure to failure without losing your enthusiasm."[53]

The quotation above resonates with me a lot today. I believe it doesn't just apply to entrepreneurs, but to everyone. Every day, each of us face difficult situations. The question that we need to ask ourselves is: do we let these events break us or do we let them define us? I believe it's the difficult times that build character and make us who we are today. I would not be the person that I am had I not chased my dreams and pushed forward. We should all embrace challenges as part of growing and learning. There are still many chapters yet to be written and I'm looking forward to seeing where the next journey takes me.

---

[53] Often attributed to Winston Churchill:
https://quoteinvestigator.com/2014/06/28/success/

The Noosphere and You

By Brian Roemmele

Introduction

In early 1922, Pierre Teilhard de Chardin introduced his idea of the noosphere[54], from the Greek νοῦς (nous "mind") and σφαῖρα (sphaira "sphere"), in lexical analogy to "atmosphere" and "biosphere": the third in a succession of phases of development of the Earth, after the geosphere (inanimate matter) and the biosphere (biological life). Just as the emergence of life fundamentally transformed the geosphere, the emergence of human cognition fundamentally transforms the biosphere. A mind encircling the world. The Internet was the first obvious step in this direction, however it is not quite a noosphere, it requires something more.

In 1963 Dr. Joseph Carl Robnett Licklider, the director at the U.S. Department of Defense Advanced Research Projects Agency (ARPA), wrote a memo about an "Intergalactic Computer Network"[55]. His research led, among other things, to the ARPAnet, the direct predecessor to the Internet. The "Intergalactic Computer Network" is aligned with the concept of the noosphere, manifested in a material system.

The inventor of the World Wide Web was Tim Berners-Lee. His HyperText Protocol described a web of hypertext documents to be viewed by browsers using client-server architecture. But Lee saw a potential problem early on:

---

[54] https://en.wikipedia.org/wiki/Noosphere

[55] http://www.kurzweilai.net/memorandum-for-members-and- affiliates-of-the-intergalactic-computer-network

> "Most of the Web's content today is designed for humans to read, not for computer programs to manipulate meaningfully. Computers can adeptly parse Web pages for layout and routine processing—here a header, there a link to another page—but in general, computers have no reliable way to process the semantics"[56]

Lee knew in 2001 even the 2017 Google knowledge graph would be very limited. He proposed a new organization:

> "A solution to this problem is provided by the third basic component of the Semantic Web, collections of information called ontologies. In philosophy, an ontology is a theory about the nature of existence, of what types of things exist. [...] Artificial intelligence and Web researchers have co-opted the term for their own jargon, and for them an ontology is a document or file that formally defines the relations among terms. The most typical kind of ontology for the Web has a taxonomy and a set of inference rules."

---

[56]http://www.foxnews.com/tech/2016/12/28/amazon-alexa-data-wanted-in-murder-investigation.html

As of 2012 Google had over 570 million objects and more than 18 billion facts about and relationships between different objects that are used to understand the meaning of the keywords entered for the search. The Semantic Web is the foundation on the path from data to wisdom.

The Journey From Data To Wisdom

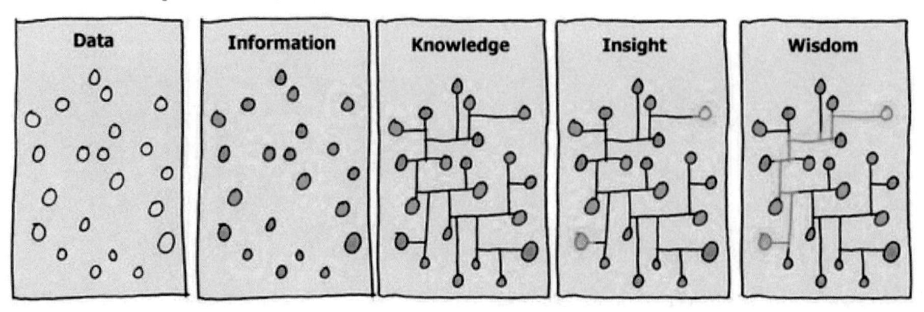

What we truly seek is knowledge with insight and wisdom – not raw data.

Data is raw. It simply exists and has no significance beyond its existence (in and of itself). It can exist in any form, usable or not. It does not have meaning of itself.

Information is processed data whereas knowledge is information that is modeled to be useful.

Knowledge is the appropriate collection of information, such that its intent is to be useful. When someone "memorizes" information, then they have amassed knowledge. This knowledge has useful meaning to them, but it does not provide for, in and of itself, an integration such as would infer further knowledge.

Insight- Insight is cognitive and analytical. It is the process by which we can take knowledge and synthesize new knowledge from the previously held knowledge. Understanding can build upon currently held information, knowledge and understanding itself.

Wisdom- Wisdom calls upon all the previous levels of consciousness, and specifically upon special types of human programming of moral, ethical codes, etc. It tries to give us understanding about which there has previously been no understanding, and in doing so, goes far beyond understanding itself. Wisdom is the process by which we also discern, or judge, between right and wrong, good and bad.

The journey of the entire computer age right up to the smartphone is to gain insights and wisdom. Up until just the last few years all we had was data, information and some knowledge.

Voice First Systems: Personal Assistants, Intelligent Assistants, Intelligent Agents

Voice First systems such as Alexa, Siri, or Cortana, will allow us to just ask a question or issue a command. It all started with simple questions or commands like "What time is it?", "What is the weather like?", or "Set a timer". With time, the types of questions will expand. Some call these early Voice First systems – erroneously – "Intelligent Assistants" or worse, "Personal Assistants". I present a more nuanced definition of what we current have:

Intelligent Agent- Acts on behalf of Intelligent Assistants and Personal Assistants. Can be short lived for just a single task for one time, like finding the time of a movie or a list of local events. They can be long lived and do this on a regular cycle. Intelligent Agent have no, or very limited information about the user and thus can generally be assumed to be like search terms used during a series of Google searches.
Intelligent Assistant- Acts on behalf of the Personal Assistant or the human user. This system forms a software (and sometimes a hardware) layer between the actual Voice First device and Voice OS and the user. It can be a layer on top of the Voice First system via the OS or voice apps. Intelligent Assistants should not hold any private, confidential or personalized information on the user. It can however hold context of the user.
Personal Assistant- Acts on behalf of the human user. Personal Assistants will hold the most personal data on behalf of the user, every aspect of the user's life. This includes health, financial, location, shopping, email, phone, web surfing data, and more. The basis of

which is to mediate interactions with other Personal Assistants, Personal Agents, Intelligent Agents and Voice First systems.

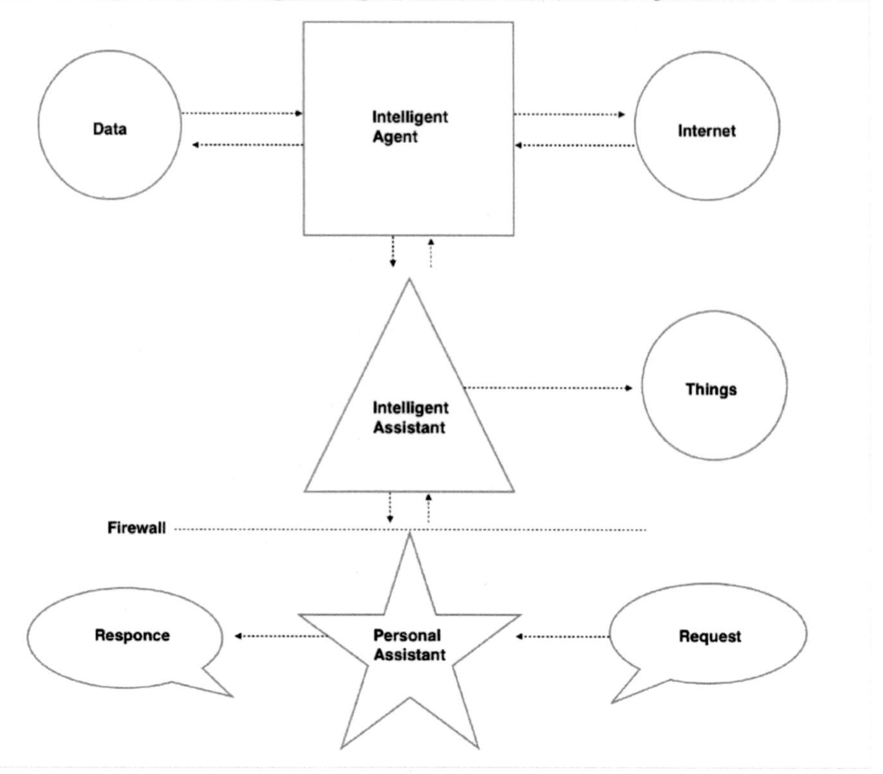

Why Not Just One "Thing"?

We will have *many*, if not countless Intelligent Agents, some operated on our behalf by companies. We will have a *few* Intelligent Assistants that we operate directly. Intelligent Assistants will interact with Intelligent Agents and the Voice First hardware or Voice OS. In some cases, the Voice First hardware company will supply some of these agents. Intelligent Assistants will at times hold contextual information about you and perhaps simple payment information. In both cases it will be situational and temporary.

You will likely only invest into *one* Personal Assistant as it represents a big investment and perhaps something worthy of the highest security in your life. The Personal Assistant will be one of the most important stores of personal information you will ever have in a computer. It will allow for the ability of the AI to know you and to understand your context. For a Personal Assistant to be of the greatest use, it will need to know the context about you to the deepest detail.

Who Owns The Personal Assistant And Where Does It Live?

My Personal Assistants (I use five in my research) have been following me on a steady basis for about three years. One system has followed me intermittently for over ten years. None of these systems are actively connected to a network and when they are it is discreetly controlled for seconds with full monitoring and verification of outgoing data. It is behind a firewall that changes virtual location ever few seconds. Finally, there is triple encryption between the Personal Assistant and the Intelligent Assistant.

There is a clear need for security. In the history of human existence there has never been a more pervasive potential intrusion into our life than the creation of a true Personal Assistant. You would never allow or agree to consciously have one central intrusive database of this nature living in a cloud computer.

It should be clear why calling Siri and Alexa a Personal Assistant or even an Intelligent Assistant under the definitions I use woefully misses the mark. It should also be very clear that once one understands just how much central data that will be accumulated on you and the greater circle of relationships one may have, there would be no desire to commit to a Google Gmail model of "You get to see my private life in hopes that you keep it protected and anonymous".

Privacy is the greater debate of our epoch and the Personal Assistant privacy debate will be one of the most challenging. Consider the Personal Assistant that has the sum total of 25 years of your life? Fifty years of your life? Access to this data would only require a question "Where was Bill on January 10th, 2025 at 11:22pm?"

When you pass away, who gains access to your Personal Assistant? Your significant other? Your children? The State? A museum? An organ donor? Would you want after 100 years of following you since the day you were born, what would be a close representation of how you would answer just about any question to remain? What are the upside effects? What are the downside effects? Consider a tragic accident that takes away a Dad from a 10 year old boy. Would you want your son to be able to ask at a crucial point in his life "Dad, what would you do?" Does it help the future or does it hurt the future?

Our Personal Assistant AI will at some point so closely represent how we would react to almost any situation, one may argue that in itself it is a sentient representation of ourselves...

Why We Really Want A Personal Assistant

With the aspects and implication of privacy of Personal Assistants, what are the upsides? Our Personal Assistants will become orders of magnitude more important to us than any Google search or old email. This is because this data and information is already not only considered, but also built into neurons that will constantly grow with new data and information over time. Google, Bing and others have created a vast, almost incompressible trove of data and information. As we presented above, Data->Information->Knowledge->Insights->Wisdom chains require the user to do most of the heavy lifting to draw knowledge and insights and perhaps some wisdom. Our Personal Assistants will have the context and have cognition that will emulate strengths of the human brain, including parallel processing and associative memory to develop context-based hypotheses on us as we relate to the world. If we can use it, see it, hear it, our Personal Assistant will not need any customer API permissions to access search results, social networks, GPS data, phone logs, banking sites, investment sights we would give it implicit and explicit permissions.

Personal Assistants, as the natural successor of any form of computing witnessed in the past, will be the ultimate Intelligence Amplifiers for our lives.

We Will Build The Early Noosphere In This Decade

We will reach the fundamental tipping point to de Chardin's noosphere in the next 25 years when the network effect of billions of Personal Assistants, with permission, will establish limited connections between each other and form a synergy that is nearly impossible to imagine today. Quite unlike the dystopia we see in Hollywood movies and pulp Science Fiction, and much like the highly decentralized of Licklider's "Intergalactic Computer Network", it will be decentralized and not a singularity of Skynet in the Terminator movies of the 1980s. de Chardin, Licklider, and Berners-Lee all formed the foundation for the likely future of our Personal Assistants, Intelligent Assistants and Intelligent Agents. And all of this is taking place now, not decades away.

We humans have gathered data and information for over 100,000 years. From cave painting, cuneiform ceramics across the ages to books in a library, binding together people who have never met each other over long spans of time and space. We humans created myths and allegories to help explain our journeys through time. We all long to share who we are and learn about our world. We all long to share our stories. The future is coming at us if we welcome it or try to ignore it. Billions of Personal Assistants interacting through Intelligent Agents billions of times per second will change our every single aspect of our lives. Nothing will be untouched by this noosphere and it will make the Internet look like ancient telegraph lines.

Perhaps at some point all of the sum total of data from our Personal Assistants would create a simulated brain of the over 20 billions of volumes of information we store in a lifetime. Maybe this is the arc of our evolution where we demonstrate that nature wastes nothing. Where the sum total of our experiences is stored and ready for the future to access. Perhaps by being able to see just how much we are alike, the differences that have defined our conflicts may fundamentally be transformed. Not a utopia, just better knowing and wisdom.

Our Voice will be the way we interact with the noosphere and it will be a Voice from the noosphere interacting with us.

Conversational Commerce: From Intelligent Assistants to Bots & Back

By Dan Miller

Opus Research coined the term "Conversational Commerce" earlier this century to describe the integration of automated speech recognition and text-to-speech rendering with business rules, logic and databases that improved person-to-person, person-to-machine and machine-to-machine conversations. Today Facebook, Slack, KIK, WeChat and others appropriated the term to mean "bots on a messaging platform".

There may be large opportunities around non-conversational bots. However, in this piece we'll document the application areas and use cases that leverage the Smart User Interface, Natural Language Understanding and Machine Learning, to support Virtual Personal Assistants, Advisors and Enterprise Intelligent Assistants.

Most Popular Use Cases and Vertical Markets

The longest-standing implementations were little more than a natural language front-end to a static data set, enabling Web site visitors to use their own words to get answers from FAQs, product or service descriptions or other marketing collateral. Telecommunications, food and beverage, travel and entertainment, banking, finance and online retailers were the first large enterprise companies to integrate Intelligent Assistants into their self-service infrastructure for customer care and marketing.

The vast majority of the first implementations were not speech-enabled. But recently– led by the innovators that include Hyatt Hotels, FedEx, US Airways, Dominoes and USAA – voice-based Enterprise Intelligent Assistants have made their appearance. They are now joined by speech-enabled virtual assistants made accessible through microphone-equipped devices as they proliferate inside homes, cars as well as a growing number of kiosks in bank branches, shopping malls and retail outlets.

Speech-enabled assistants present themselves as "skills" for people who converse with Amazon's Alexa or "actions" that can be taken through Google Home Assistant. Starting with built-in functionality such as weather, timers/alarms, music, shopping and trivia, these skills later expanded into daily necessities like ordering a pizza, coffee, car service, movie tickets or a restaurant reservation. The things you can do after saying "Alexa", "Echo", "Computer", "OK Google" or other wake-up words keeps growing. The comfort level in invoking those words is growing in parallel.

Sizing up the Market

2016 witnessed explosive growth in enterprise spending on licenses, services and platforms that support Intelligent Assistance. Investment was roughly $750 million in 2015, which was more than double the $350 million originally forecasted by Opus Research in 2013. At this rate, Opus Research foresees explosive growth of the industry poised to blast through $1 billion in 2016, on the way to $4.5 billion globally by 2021.

Figure 1: Forecast Spending on Enterprise Intelligent Assistants

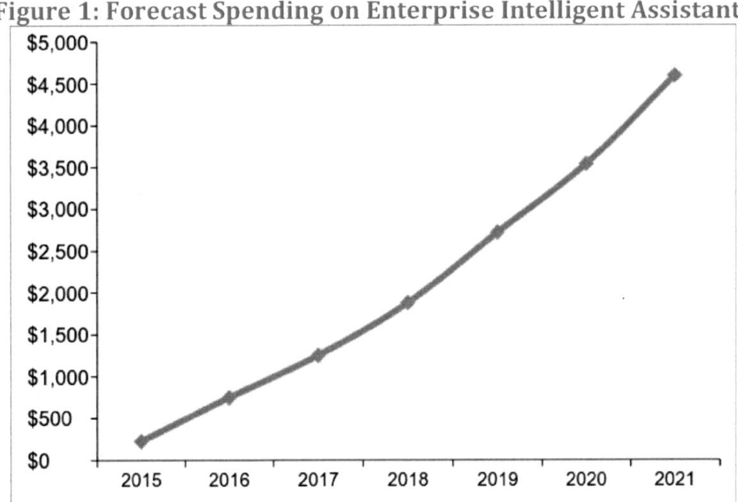

Source: Opus Research (2017)

The ranks of firms offering “Enterprise Intelligent Assistants” platforms and services have grown accordingly. In a report entitled “The Decision Makers’ Guide to Enterprise Intelligent Assistants” Opus Research evaluated offerings and implementations of twenty-eight firms whose 1,200 enterprise customers are responsible for over 2,500 implementations – of bots and virtual agents to support both chat and phone-based conversations.

Recognizing the Intelligent Assistance Landscape

Without benefit of a common vocabulary and shared landscape for Intelligent Assistance, those trying to keep up with technological advancement and competitive offers run the risk of trying to see beyond your own headlights. Below is an illustration and description of the Intelligent Assistance and Bot Technology Landscape.

Figure 2: Technology Stack for Intelligent Assistants and Bots

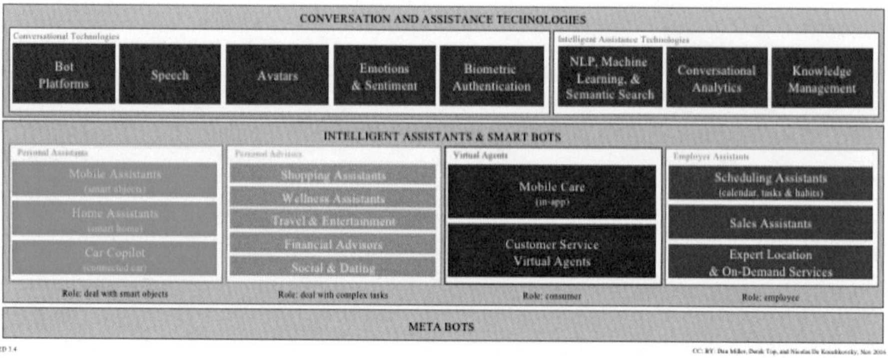

CONVERSATION AND ASSISTANCE TECHNOLOGIES

Conversational Technologies: Bot Platforms | Speech | Avatars | Emotions & Sentiment | Biometric Authentication

Intelligent Assistance Technologies: NLP, Machine Learning, & Semantic Search | Conversational Analytics | Knowledge Management

INTELLIGENT ASSISTANTS & SMART BOTS

Personal Assistants: Mobile Assistants (smart objects) | Home Assistants (smart home) | Car Copilot (connected car) — Role: deal with smart objects

Personal Advisors: Shopping Assistants | Wellness Assistants | Travel & Entertainment | Financial Advisors | Social & Dating — Role: deal with complex tasks

Virtual Agents: Mobile Care (in-app) | Customer Service Virtual Agents — Role: consumer

Employee Assistants: Scheduling Assistants (calendar, tasks & habits) | Sales Assistants | Expert Location & On-Demand Services — Role: employee

META BOTS

Conversational Technologies: The Front End of a "Smart UI"

The top row of the Intelligent Assistants Landscape depicts the functional components of IA platforms. It is divided into two rectangles: Conversational User Interface (UI) Technologies and Intelligent Assistance Technologies. These listings incorporate the comprehensive set of capabilities that should be considered when evaluating candidates for IA platforms or solutions.

Figure 3: The Conversational Front End

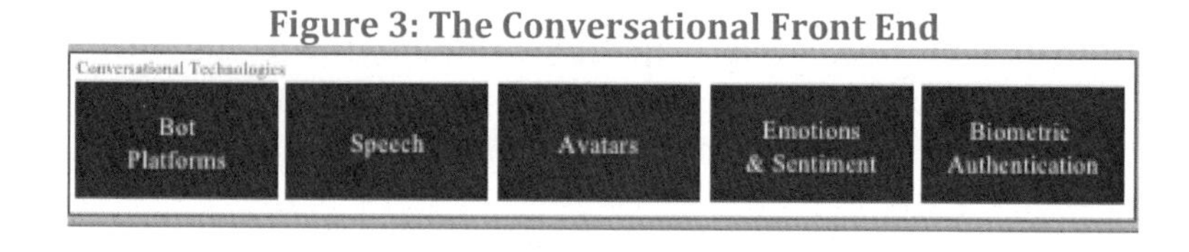

The major technology elements that comprise the Smart UI today, and for years to come are as follows:

- **Bot Platforms** – "Conversational Commerce" refers largely to text-based communications over messaging platforms. This is a high-growth/high-potential channel.

- **Speech** – With recognition accuracy exceeding 90% and extremely human-like spoken output, Opus Research expects to see growing acceptance of voice as a universal interface for intelligent assistants on phones, smart home devices, consumer electronics, automobiles and retail kiosks.

- **Avatars** – Visual representations of virtual agents have long been a part of the presentation layer of conversational technologies. End-users like to anthropomorphize their intelligent assistants and the technologies to render animated beings, both life-like and phantasmagorical are steadily improving, making the conversational UI more immersive and engaging.

- **Emotion & Sentiment** – "Context is King" is one of the watchwords of effective digital commerce. In addition to the basics of location, purchase history, age, income and other status indicators, detecting and responding to the emotional state of an IA's client or customer is becoming vitally important. Judging platforms by their ability to show empathy and understanding will be more than a differentiator.

- **Biometric Authentication** – Newly added to the latest version of the IA Landscape, this category considers the ability to recognize and authenticate an end-user based on their touch (fingerprint), something they say or how they say, or other means of authentication. This category is emerging as a crucial component to providing a highly personalized, trusted link between individuals and their IAs.

Not all of the above categories exhibit the same level of maturity or general availability.

Intelligent Assistance Technologies

The rectangle on the right side of the top row is where the magic happens. These resources provide brands with the ability to recognize an individual's intent quickly and find the best answer to a question or the best action to suggest in response to a command.

Figure 4: Enterprise Intelligent Assistance Components

- **Natural Language Processing, Machine Learning and Semantic Search** – These computing and database management resources can be either "rules based" or employ deep neural networking to teach themselves how to recognize the meaning of all the data they course through from textbooks, journals, newspapers, and elsewhere. This data is then used to create a model that matches with the requests from individual customers, prospects, clients or members.

- **Conversational Analytics** – Many large, sophisticated enterprises and brands have already invested computing resources that monitor customer interactions (call center recordings, chat transcripts, etc.) to detect patterns that correlate with successful interactions or detect when companies must take remedial action. Applying these lenses to real-time conversations will make IAs more efficient and predictive in what they are offering.

- **Knowledge Management** – As IAs proliferate and expand their capabilities and areas of expertise, it is becoming increasingly evident that each company's knowledge base (be it product literature, CRM records, chat transcripts, or FAQs) is the raw material that makes Intelligent Assistance possible.

Great attention should be paid to a potential solution provider's methods for ingesting and organizing data (both structured and unstructured) from a variety of sources. In addition to FAQs, the material to be included should include chat transcripts, call recordings and transcripts, product literature and manuals and threads from social platforms, all with an eye to those that represent "best answers" and successful resolution.

Intelligent Assistance, therefore, is the result of the artificial intelligence resources making the best match between the information it has ingested and the requests coming from individual customers or clients. For the purpose of understanding the context of a call, it will be advantageous to integrate information about the individual making contact, preferably in real time. It is routine for companies to know a customer's location, the device they are using and its capabilities, information on recent activity (especially regarding the company's website or activity within the mobile app) and their "status" as a preferred customer or member of the company's loyalty program.

Merging the Parallel Universes

Customer care professionals have a long history of rapidly recognizing intent and responding to it in a way that is both pleasing to the customer and cost-effective for the company. They brought natural language understanding (NLU) to the intelligent assistance platforms that power speech-enabled voice response units, virtual chat agents, and text-based prompts that appear in "screen-pops" on the screens of live customer service reps.

Bot developers cut their teeth on RESTful APIs, agile development, scrums, and hackathons. The fruits of their labor are "simple" messaging bots that often take a one-and-done approach to task completion. They order a pizza, schedule a meeting, or deliver a news report, then move on. They will benefit from speech-enablement but, as a norm, "conversation" is not necessary.

Today, the two communities blithely coexist, but live in parallel universes. The bot makers think the customer care professionals are making things too complicated, while those customer care pros see the botsters as an existential threat, luring customers from their captive CRM systems into the freer-feeling (though equally captive) world of social media. Popularization of Alexa, Google Assistant and all assistants that follow are destined to end the schism and bridge the chasm.

Just Talk - a Machine Will Help you!

By Detlev Artelt

1,000 years ago, we humans learned to speak. This basic human-to-human communication allows us to understand each other today at an extremely detailed level. Whenever two people meet, they agree shortly on a common language. Then, a conversation on any topic can take place.

People communicate with each other all over the world, and have designed a behavior that leads to a successful communication. We created "conversation" - we speak, ask a question, answer a question with answer or another question, and so on. This type of communication has worked quite beautifully for many centuries, and is the basis of today's culture.

Now computers join the conversation

Some decades ago, computers joined our lives, and we were forced to give them machine language commands for certain tasks. But unfortunately there was no audible language to communicate with machines; they only understood characters in the form of text. So we had to laboriously press keys to input our thoughts, to deliver content and commands to the machine. With speech technology, this process could be so much more pleasant.

Soon, people began to wish that they could also speak to their machines, and have them speak back. Many decades of research and development passed until we came to a proper conclusion. [So I sit here and dictate this text to my machine, via headset].

Parallel to speech recognition research and development, the Internet has evolved for the last 20 years and only recently we start to talk about a complete digitization of our lives. Also here we communicate mainly by keys with a machine that sends it to another machine to present it then a human. But unfortunately, all this happens in the form of writing and not in the form of voice or audio. We're used to operating the browser via the keyboard, chatting via WhatsApp (or other Messenger services) via text, and rarely using spoken language. Lately, we see the people use the messaging services to send spoken messages to each other, which is at least a first step into a direction where we use more audio.

But now, there is significant change happening in the computer and speech industry -- we will talk more to our machines, listen to them talk back, and learn a new safety through language. This new language technology is available everywhere around us.

Speech revolution

Unfortunately, however, man is a creature of habit and has become accustomed to operating the machine using keys. It is always something special, even something unknown, when we talk to a machine. But now high quality language technology (in many languages) is available worldwide and we must take advantage of this efficiency. A few years ago Siri showed on the iPhone what is generally possible. This provided for an increased acceptance of language phrases. Amazon is now well known for the Echo, and Google Home is available on the market as well. This is only the beginning of a new revolution in the operation and control of our entire life - if we decide to engage with it. But do we have a choice at all?

Examples of current language technology

The simple input and switching off of lights in the home or in the office is fundamentally changed by the home automation. But most still turn individual lights and have less control of desired light scenes. It would be so easy if we would just talk when leaving the House "**Light off**". Of course, for convenience there will be functions allowing (for example) a whole floor to be controlled. The command "**bring the lights on the second floor to 30%**" replaces the presses and turns of many switches -- this significant increase in efficiency is obvious. With all the latest LED technology we get colored lights available, and speech is a far better for choosing the color you desire by saying "**Switch the lights to orange in the living room**". Similar functions can also be used in a computer for a conference room, to control the lighting functions of the room perhaps between "**Teamwork**" and "**Presentation**" mode. If the mode automatically changes, this is a convenience and allows for better work efficiency.

The new customer service

Expanding these basic speech functions with some more business knowledge might work very well in customer service. It could especially help on possible ease of operation for the caller, and thus to ensure a quick success in the interaction. That data available to the company can be used to enrich the conversation, so they can faster identify whether the user has just submitted an order, has already called on the previous day in the service, or that they already communicated the identical concern using other channels such as email or chat. Special security allows the use of biometric characteristics of the voice that can be used to identify people. Every human being has over 1000 different characteristics in his voice or the neck nasopharynx, which generates the voice; voice biometrics is able to determine these parameters to identify and authenticate the user.

Analysis of the language

The analysis of the recorded conversations from customer service hotlines currently allows a look at the content of the discussion between caller and client on the phone. So the words and phrases they spoke will be analyzed. The reaction from the agent to whether the caller said might be used for training the agent. Words such as order, notice, lawyer or other theme-related terms may act as triggers. Due to the compaction of the words obtained in the analysis a look at the content of the communications can be the first automated throw and completely new observation points and conclusions follow. Often, certain words or even mild cursing such as "stupid" can slip out once the agents or the caller, but it should remain an exception. So far the management has been confident that these words are not used, but through an analysis using language technology the defined taboo words can be find and assign specific persons or numbers.

Big data and voice

The real-time analysis on the emotions can go further. Speech technology with enough calculation power is able to work in real time on what and how somebody said a phrase or sentence. Current language technology is capable of different emotions such as stress and tension. Such feelings or other non-verbal cues in a conversation are recognized sufficiently, so this can lead to better success in communication. The agent can support such emotion detection itself and reflect how and where he should engage maybe on some other interaction in the conversation, or even better control the conversation by early objections.

The real-time analysis shows at a glance how emotionally involved, friendly and committed is the customer in the course of the conversation, and also gives information about the call quality created by the agent. This data can be used very well for training purposes and so agents on the different areas such as customer service, sales, can optimally on all conceivable situations prepare fault hotline or complaints management. It also helps the agent to understand that he might have a break for a few minutes to relax as he might deliver a better service after having a coffee or a tea.

Get in-car information

There are huge amounts of information available in the car: the planned travel route, the traffic situation, the media collection in the vehicle, or information about the vehicle itself. All of this information about the driver can be gathered without distraction to evaluate and apply for his needs. Of course, it can work in this environment particularly well by voice automation as the driver won't need to press all that different button, menu items or control knobs.

On the other hand, many of the current car solutions use the cloud as the speech understanding part -- and they only work if there is an internet connection. Because it unfortunately often is not the case, one needs to be careful when designing solutions using language technology in the vehicle.

Data security

Wherever the language technology is used, there is a recording of the personal audio signal, your own voice, and often this recording is transmitted via the internet. Then that recording is analyzed, and the content is recognized by a machine in a data center in any country. The result will be send back to the user but perhaps left a lot of metadata for tracking the content or even the user. At this stage a provider or even a country might be able to collect many information about the individual.

At this point it is just thinking what requirements for the security of data and personal content are attached. In Europe and especially in Germany the use of personal data is a huge issue, because we humans are as a result more and more transparent for the major brands.

The personal dictation of a letter for a lawyer, which was edited in the cloud solution -- who is it up to to evaluate its availability according to the privacy policy of the provider? This is where there needs to be clarity to decide how and where the personal data is used.

The way for everyday use

For several years now, there have been personal assistants; Apple's Siri and Google Assistant have gained some fans and friends. But the breakthrough in the mass market and the very obvious use of language for the control of machines has still not reached the highest level of acceptance.

More recently, there has been Alexa from Amazon, better versions of Cortana from Microsoft, and changes to Siri by Apple. Another big event occurred in the industry when Samsung bought Viv (successor of Siri) and of course, there are other solutions from Google Now. Therefore, there are now a lot of voice applications that all have AI functions on board. Now, there are more and more machines with high-accuracy speech recognition and natural language understanding. So we can expect a variety of options for the operation of machines available -- at least on the manufacturing side.

But what about the users?

The coffee machine that understands the command "**A latte with triple espresso**" for the individual task must not be an individual closed solution, but should be part of the entire automation system. An interface for all and not 10 different systems are surely beneficial from the perspective of the user. So then the TV can be perfectly controlled with "**Switch at 18:00 to CNN"** without many buttons to press or even a programming to enter.

Language technology is constantly around us, and we are able to use it every day. But we need to start using it way more in order to drive full acceptance and understanding of this valuable asset.

The Future

Just try it, play with voice. Download the latest app for you smartphone and give it a try. If you are on an iPhone or a Mac, just try to dictate an article as all speech recognition is integrated. It will slowly change your life to a more convenient way to communicate with your surroundings. Start with a small device in your house and maybe not with the fully equipped house for the cost of several thousand dollars or euros.

Whenever you look for your next car, check what kind of speech technology is integrated and double-check also if it works whenever you are off the main tracks, or not connected to the internet.

But also start carefully thinking about your data. Sure, you might tell Twitter and Facebook everything about your life already, but you have to decide yourself how transparent you'd like to be for all others.

With all the upcoming artificial intelligence solutions, the customer service will reach a new level. Try to use these services and learn how you might get a faster service.

Most important, speech is on the rise. We will see way more easy-to-use technology, and it will be fun to use all of it. So "just talk and a machine will help you!"

When Liberal Arts meet Technology

A Computational Linguist's view of Conversational User Interfaces and their Impact on Customer Service

By Tobias Goebel

Steve Jobs once said[57]: "[...] technology alone is not enough—it's technology married with liberal arts [...] that yields us the results that make our heart sing." What might sound corny to some simply reflects the truth for those who care about User Experience Design.

When it comes to Conversational User Interfaces such as chatbots on Facebook Messenger, Slack, or even SMS, or voice UIs on Amazon Echo or Google Home, we seem to be making the same mistake that we already made with similar systems in the past: letting the "developer" be the same person as the "designer" of the man-machine interface. If we want the new generation of CUIs to be successful, we need *you,* the linguists, the writers, the designers, the psychologists, and the liberal arts majors. Humans ended up at the top of the food chain for a number of reasons, but chief among them our ability to speak, i.e. communicate efficiently. Language lets us collaborate towards a common goal most effectively. Each of us brings unique skills, and it is by putting them all together that we can accomplish great things.

This essay explores what happens when you build technology without considering the human factor.

---

57 http://www.newyorker.com/news/news-desk/steve-jobs-technology-alone-is-not-enough

Our Thinking Cannot Rely on Logic Alone

When engineers go through their professional training, they learn to *solve problems*. As part of that, they learn to decompose complex matters or machinery into smaller components. Their thinking is guided by mathematical precision and Boolean logic (after Englishman George Boole, a 19th century philosopher), which is the foundation of our binary computer systems. Something is either black or white, true or false, on or off, 1 or 0. There is no "maybe" in traditional engineering. Over time, being constantly exposed to this type of thinking, some engineers (be they mechanical, electrical, software, civil engineers, or a combination thereof) tend to let these principles permeate into their personal lives. They apply logic to everything, but in "real life", especially when it comes to human behavior, not everything follows the rules of Boolean logic.

Contrast that with liberal arts or humanities majors. They, too, learn to solve problems – yet the problems *they* solve are grounded in human nature. They learn the "tools of society", such as language, literature, arts, psychology, religion, social sciences. Their thinking is not guided by the rules of Boolean logic. Why? Because humans aren't. There is no Black and White in human nature. We are a species of nuances.

Sometimes, you find academic disciplines that marry these two worlds. I myself have chosen to go down such a path. After finishing high school in Germany in 1997, I decided to immerse myself into a field called *computational linguistics* for 6 years – an explicitly interdisciplinary subject that connects the world of computer programming (engineering!) to the world of human language and the study of syntax, semantics, morphology, pragmatics (dialog management), as well as the basics of phonetics and phonology. This academic field combined my interests in math, computer programming, and language. Ever since I worked in this field, I have remained an interdisciplinary worker till today. I find joy in bringing different worlds together, in connecting people of different backgrounds. Sometimes you even have to "mediate" between these groups!

The Voice User Interface Enters the Mass Market

Fast forward 18 years. Amazon releases a product that might one day be looked at the same way we look at the personal computer or the iPhone today: the Amazon Echo. As a matter of fact, I consider the Echo the most impressive device in consumer electronics since the iPhone came out in 2007. And the impact of the iPhone on society as a whole cannot be overestimated. What makes the Echo so unique? It is the first time a major technology company has the audacity to release a product with only a voice interface, and nothing else. Essentially no tactile interface, no graphical interface (if you ignore the companion app); just a Voice User Interface (VUI) to access information or conduct transactions.

Amazon understood that the technology itself will be of no use, will not find a big audience if there isn't an application that improves our lives. Only by appreciating that a superior mechanical product (an array of microphones arranged in ring form, tuned for far-field recognition, and speech recognition algorithms optimized for indoor, at-home use) must be embedded into a setting that fits into people's lives were they able to land a hit: the kitchen and living room seem to be the most favorite spots for the Echo.

A VUI as the only interface to the Echo means that we must use our voice to operate it. And everyday "spoken" language (even written language shares the style of oral language when it is conversational) is still the most natural way for us to ask for information or get stuff done – that will never change. We start learning the phonetics of a language while still in our mothers' womb, learn the "voice API" to other humans by the age of two, complete language acquisition roughly at four years old. From then on it is vocabulary and style; and irony, sarcasm, and all the other fun forms of human (mis-)communication.

One area of application for the Amazon Echo is customer service; soon after the product's release, some forward-thinking companies started releasing skills that let consumers access business data through voice. That makes sense, as everything in customer service starts with a question or need in our minds, formulated using words:

- *"What is the status of my claim?"*
- *"When can I expect my order?"*
- *"Why did I get charged a $15 processing fee?"*
- *"How can I reset my password?"*
- *"I need to pay my credit card bill"*

Contrast that with a GUI (Graphical User Interface), and you will realize that in order to get stuff done on GUIs such as websites or mobile apps, you have to transform the question you have in your mind to a sequence of menu item selections. That is an additional mental hurdle. Not so with VUIs: just ask your question. The VUI (or Conversational UIs in general, which includes messaging/texting/chat) can shorten the path from your question to the answer like no other UI can.

How GUIs Differ from VUIs or CUIs

Conversational UIs differ from GUIs in one critical way: user input cannot be controlled. But when there are no buttons to press, fields to fill in, or any other guidance by the UI itself, just the "blinking cursor", humans can get creative!

A while ago, I tried out a virtual assistant built into a retailer's website. The assistant engaged me in a conversation about jackets. It began with the prompt, *"Where and when will you be using this jacket?"* I typed, *"I need one for a skiing trip to Mass."* It then went on to ask if this was for a man or a woman. Eventually, here is how the system interpreted my responses:

Gender: "male"
Designed for: "skiing, cold, winter"
Travel destination: "one"

The "artificial intelligence" misunderstood *one* as the destination of my trip, versus the lovely state of Massachusetts! Now you might say that I didn't answer the question, since it asked specifically *"where and when will you be using this jacket."* But my interaction illustrates the very challenges of human language: it is highly ambiguous. And often, people won't respond how you expect (and want) them to. One sentence can mean many different things, depending on the context. Unfortunately, the engineers who have designed many of the text- or voice-based interfaces in the past, have studied the world of programming languages — which are the direct opposite of natural languages: no ambiguity of meaning, instead clear and straightforward syntax and semantics. One statement in a programming language means one thing, and one thing only.

We've seen this before in the world of Interactive Voice Response (IVR) systems. Too many applications were built by engineers, who know very little about how language works as a means to communicate between human beings, let alone how to make machines use it. Here's a classical example of an IVR system giving stock quotes:

Apple is at 94 dollars and 2 cents, down two dollars 58. Would you like to hear another one?

Microsoft please

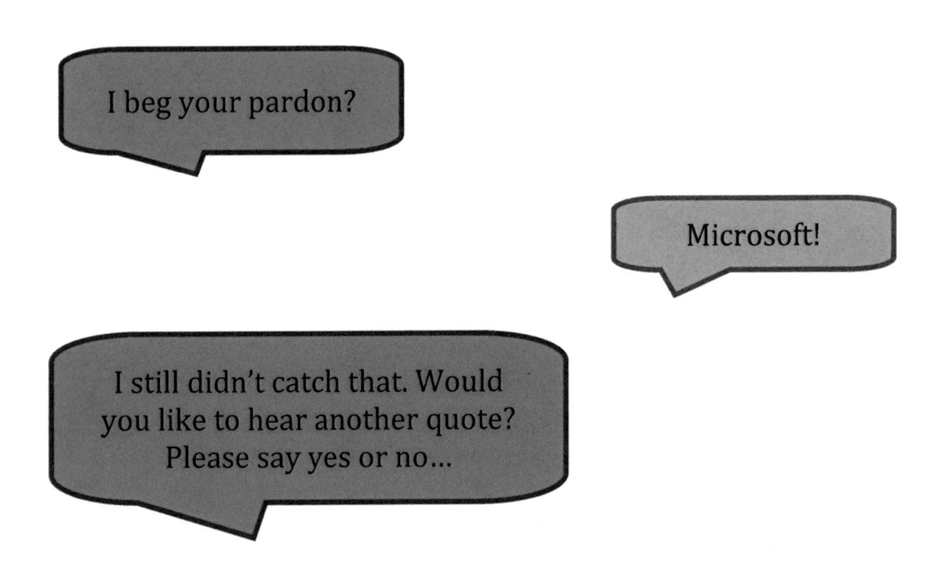

"Microsoft" is a perfectly acceptable answer to this question, albeit not what the engineer had in mind when he or she programmed the system for a "*yes*" or "*no*" answer.

The trick lies in carefully thinking through the entire spectrum of possible answers when designing a question in the system: what could the user say as a response to my prompt? Engineers tend to take questions literally, whereas non-engineers often do not. And politicians have mastered the art of taking a question not literally... (Should a design principle for CUIs be: design with a politician as your end user in mind?)

Is There a Linguist in the House?

Flocks of developers are jumping on this fascinating new thing called Conversational UIs, only to build chatbots or voice skills that are mediocre at best, or counterproductive at worst. My concern is that we will quickly ruin this emerging field and make early users turn away and dump this new form of UI, telling others: "*doesn't work, stick to the app*."

At a recent Amazon Echo developer and partner event, the VP of Digital Products of a major US bank that was among the first to release a skill for the Echo told the audience that they hired a former storyboard writer from computer animation film studio Pixar (!) to help with their emerging conversational UIs. *That* is exactly the type of skill we need: Writers, actors, authors, general linguists, computational linguists, psychologists, and human-machine interface specialists. People trained or experienced in these fields understand the complexity of language and communication and won't repeat the mistakes of the past, given they are empowered to do so and occasionally say "no" to a customer who thinks they know better because they, too, speak English. The communication professionals will create voice experiences that truly help us with our stressful lives, fit into our lives, and build solutions that make things easier, not harder. Faster, not slower. Mid- to long-term, Conversational UIs have the potential to fundamentally change how we work, do business, live our lives. Let's try to get it right from the start. Let's involve the people that can help us.

Acknowledgements

This book is a "labor of love" as one of the authors so eloquently stated. It would be impossible to thank everyone who supported it along the way, but here is my humble attempt.

First, I'd like to thank my friend Ahmed Bouzid, who recruited most of the original authors. Without him, this book would not exist. And to the authors themselves, all fifteen of them, it has been an honor getting to know you more and reading your work! You are all incredibly accomplished professionals who made this experience an enjoyable one. Special thanks to Eduardo Olvera for the cover art, and to Tobias Goebel for pitching in with editing help.

I've been lucky enough to have amazing friends and mentors throughout my career that inspired me. My graduate school advisor Abeer Alwan, who accepted me into her SPAPL lab at UCLA, has always been a role model for me. I'm also indebted to fellow author and best friend Cathy Pearl. Not only did she contribute to this book, she also kept me sane throughout the process of writing and editing. To the rest of my book club "6 Books a Year": Amy Ulug, Rebecca Green, Karen Kaushansky, Elizabeth Strand Cimini and Jennifer Balogh Ghosh- I love you all for keeping me reading, laughing and eating for the past 16 years! You constantly motivate me in my writing, as well as my professional and personal life.

My work friends throughout my career, especially over the last few years, helped make this book possible. At NextEV/NIO: Brittany Repac, Alex Sempel, Anthony Nguyen, Michelle Li, Rebekah Lee, Kevin Wong, June Chan, Chris Eckert, Anthony Simonetti, Robert Miller, Brian Rink, Nick Hoppesch, Abhishek Singhal, Henry Wang, Aimi Yep, Katelyn Sandhu, Jeff Wallace and Paddy Kumar. I apologize for all those times I refused to go to lunch with you so I could work on my book instead! You brightened my days at work innumerable times, and for that I will be eternally grateful. Thanks also to my former CEO, Padma Warrior, for her support. From Amazon: Monica Farrar Miller, Garret Miller, Blair Beebe, the entire AVA team, and my long-time friend Jared Strawderman.

To my mom, you have been the best female professional role model for me from the beginning. You were able to juggle family and career without compromising excellence in both, and for that I've always admired you. Dad would be proud of both of us over the past 2 years... We are managing to live life without him, while loving and missing him all the same.

Last and most important of all, I give my genuine thanks to my husband Jason and our son Jon. I'm sorry for every time I was grumpy with you after a late night of writing and editing! I love you more than I ever thought possible.

About the Authors

Brian Roemmele

Brian Roemmele is the consummate Renaissance man. He is a scientist, researcher, analyst, connector, thinker and doer. More than anything, Brian is an observer of the world, asking questions and following his seemingly infinite curiosity to the root of an astounding variety of issues. Over the long, winding arc of his career, Brian has built and run payments and tech businesses, worked in media, including the promotion of top musicians, and explored a variety of other subjects along the way. Brian actively shares his findings and observations across fora like Forbes, Huffington Post, Newsweek, Slate, Business Insider, Daily Mail, Inc, Gizmodo, Medium, Quora (top writer for 5 years), Twitter and his own podcast, Around the Coin that surfaces everything from Bitcoin to Voice Commerce. Now he has created the Multiplex app and Multiplex Magazine, a way to stay on top of everything important in technology, payments and just about anything else.

Bruce Balentine

Consumed by a lifelong passion for the human voice and human hearing, Balentine creates interfaces that exploit speech recognition and synthesis, language, sonification, music and gesture. He has authored several books on speech technologies. Now in semi-retirement after devoting more than 30 years to theoretical and applied Human Computer Interaction (HCI), Balentine has replaced product design with a preference for education, mentoring, and workshop facilitation. He is also a composer, conductor and lecturer, holding BM and MM degrees in music composition from the University of North Texas with interdisciplinary studies in electroacoustic music and intermedia.

Cathy Pearl

Cathy Pearl is Head of Conversation Design Outreach at Google, and the author of the O'Reilly book, "Designing Voice User Interfaces". She's been designing and creating Voice User Interfaces (VUIs) for nearly 20 years and is passionate about helping everyone make the best conversational experiences possible. Previously, Cathy was VP of User Experience at Sensely, whose virtual nurse avatar helps people stay healthy. She has worked on everything from programming NASA helicopter pilot simulators to designing a conversational iPad app in which Esquire magazine's style columnist tells users what they should wear on a first date. She has an MS in Computer Science from Indiana University and a BS in Cognitive Science from UC San Diego.

Charles Jankowski

Charles Jankowski is currently Director, NLP Applications at CloudMinds Technologies, with 30 years experience in both academia and industry developing algorithms for and/or client facing applications using advanced natural language, speech recognition, speaker verification, and search technologies. From 2013—2016, he was Director of Speech and Natural Language at 22otters, building a healthcare patient engagement platform. From 2012—2013 he was with PTP, developing IVR applications. From 1998—2011, Charles was in various roles at Nuance, including IVR generalist, Speech Scientist, Manager, Senior Manager, and Director in the Professional Services team, where he managed a cross-functional team of Project Managers, VUI designers, Developers, and Speech Scientists.

Detlev Artelt

Detlev Artelt is Senior Consultant, Author and pioneer in business communications and speech technology. His latest book, "Einfach Anders Arbeiten" explains to managers the benefits we got from today's digital world to business in SME and enterprise business. As the author of the compendium voice compass, his goal is to help people to understand the benefit and the usage of speech technology in a business environment for a better communication between humans and machines. Specialties include: 25+ years knowledge in business communications, consulting in more than 150 solutions, unified communication, change management, voice platform architecture, speech recognition and speech synthetics as well as speaker verification knowledge, technical due diligence for voice solutions, call centers and speech automation. He is focusing on the new way of working by using latest technology to work in latest office infrastructure by using the new rules for employees.

Dan Miller

Dan Miller founded Opus Research in 1986 and helped define Conversational Commerce through consulting engagements and by authoring scores of reports, advisories and newsletters addressing business opportunities that reside where automated speech and natural language processing leverages Web services, mobility and enterprise software. As Director of the New Electronic Media Program at LINK Resources from 1980-1983, he helped define one of the first continuous advisory services in the information industry. He then held management positions at Atari, Warner Communications and Pacific Telesis Group (now part of AT&T). He edited and published "Telemedia News & Views," a highly-regarded monthly newsletter regarding developments in voice processing and intelligent network services. He also served as Editor-in-Chief of The Kelsey Report, where he also oversaw the launch of advisory services on local online commerce, voice & wireless commerce and global directories. Dan received his BA from Hampshire College and an MBA from Columbia University Graduate School of Business.

Deborah Dahl

Deborah Dahl designs and builds groundbreaking applications of speech and natural language technology, working with customers ranging from startups to large enterprises and government agencies, leveraging her 30 years of experience in speech and natural language technologies. She is a frequent speaker at industry conferences and has published many technical papers. Her books include *Multimodal Interaction with W3C Standards: Toward natural interfaces to everything,* published in 2016. She is on the Board of Directors of AVIOS, (Applied Voice Input Output Society). Dr. Dahl received the "Speech Luminary" award from *Speech Technology Magazine* in 2012 and 2014.

Jonathon Nostrant

Jonathon Nostrant currently works on Product and Business Development for Viv Labs (acquired by Samsung) and is working on launching Bixby 2.0 on several new Samsung devices. He was formerly Founder & CEO at ivee, which focused on designing, developing, and distributing an open & flexible voice assistant device for the connected home. Previously, Jonathon co-founded Moshi, the first speech-enabled alarm clock. He has over 10 years of voice and A.I. experience, especially when it comes to developing & distributing consumer electronics with voice.

Eduardo Olvera

Eduardo Olvera is a Senior UI Manager & Global Discipline Leader at Nuance Communications. He also advises startups on conversational UX best practices. His work on many international English, Spanish and French voice user interfaces and virtual assistants for such industry leading corporations as US Airways, Wells Fargo, USAA, FedEx, Ford, Walgreens, Telefonica Movistar, Bank of America, Fidelity, US Bank, Geico, Vanguard, Samsung, Virgin, MetroPCS, Sprint, and Sempra, give him a deep understanding of the user and business challenges of mobile, multilingual and multimodal application design, development and implementation.

Leor Grebler

Leor Grebler is co-founder and CEO of Unified Computer Intelligence Corporation (UCIC), a company dedicated to bringing voice interaction to hardware. It's initial product - Ubi – The Ubiquitous Computer – was a voice activated computing device that offered instant access to information and control of home automation devices and was the first product to offer natural environment-based voice interaction. Leor steers UCIC towards its goal of making interaction with technology more human and natural.

Lisa Falkson

Lisa Falkson has over 15 years of industry and research experience, specializing in the design, development and deployment of natural speech and multimodal interfaces. Most recently, she worked on next-generation voice user interfaces for the vehicle at NIO and CloudCar, as well as Amazon's first speech-enabled products: Fire TV, Fire Phone and Echo/Alexa. Prior to that, she focused on early iPad/iPhone interfaces at Volio, and IVR applications at Nuance Communications. Lisa holds a BS in Electrical Engineering from Stanford University, and a MS in Electrical Engineering from UCLA.

Maria Aretoulaki

Dr. Maria Aretoulaki is the CEO of DialogCONNECTION, a consultancy focusing on Speech IVR and VUI Design. She has been designing and optimizing voice self-service applications for 20+ years for clients both in Europe and the US. Her background is in Computational Linguistics, Text Summarization, Machine Translation, and Neural Networks.

Phil Shinn

Phil Shinn is principal of the IVR Design Group, which designs language interfaces. In the past Phil has been at Ttec, Voxgen, SpeechPro, Cyara, Morgan Stanley, Genesys, Bank of America, Hughes Research Labs, HeyAnita, Citibank, Unisys and Speech Systems, building speech and text apps since 1984 for dozens of companies and government entities. Phil created the open-source Voice User Interface Designer Toolkit, which has been adopted by a number of organizations. He has a Ph.D. in Linguistics (in acoustics) from Brown, wrote one of the early grammar checkers shipped with Microsoft Word, has 5 patents, and has served on the board of directors of AVIOS and AVIxD.

Sunjay Pandey

Sunjay Pandey is VP of Product Development for Capital One Investing, where he leads Product, Design, Research, Data Science, and Program Management teams focused on building financial fitness experiences and solutions for web, mobile and emerging consumer platforms. Prior to Capital One, Sunjay led Product Management for the Alexa Skills Kit (ASK), launching the SDK that now powers a growing ecosystem of developers and content creators to build voice experiences for Echo (the digital device) and Alexa (the intelligent personality). Prior to Alexa, Sunjay led Product for Amazon Simple Queue Service and Amazon Simple Workflow, both AWS services for building large-scale distributed and big data systems. Before joining Amazon, Sunjay founded and built cloud and SaaS startups for over a decade - with launches across gaming, media and advertising, enterprise collaboration, securities trading, and agile product ops. He holds a Bachelors degree in English Literature from Appalachian State University and a Masters in Management Information Technology from The University of Virginia.

Tobias Goebel

Tobias Goebel is a conversational technologist and evangelist with over 15 years of experience in the customer service and contact center technology space. He has held roles spanning engineering, consulting, presales, product management, and product marketing, and is a frequent blogger and public speaker on Customer Experience topics. He serves as VP Product Marketing and Emerging Technologies at Sparkcentral today, defining the future of conversational relationships between businesses and consumers. Prior to that, he lead Aspect Software's global chatbot strategy where he was the lead architect behind the award-winning chatbot "Margot the Wine Bot". Tobias holds degrees in computational linguistics, phonetics, and computer science from the universities of Bonn, Germany and Edinburgh, Scotland.

Wally Brill

Wally Brill is currently Head of Conversation Design Advocacy & Education at Google, where he helps Google's partners develop and maintain best practices of conversation design for the Google Assistant. Previously as Google's Senior Persona Designer he helped bring the character of The Google Assistant to life. Prior to coming to

Google, he designed persona driven, speech recognition systems for enterprises and governments from Allstate Insurance and British Airways to eBay and The U.S. Navy. Wally studied electronic music composition at The New School for Social Research.

93921012R00082

Made in the USA
Lexington, KY
19 July 2018